Hell's Kitchen and the Battle
for Urban Space

Hell's Kitchen and the Battle for Urban Space

Class Struggle and Progressive Reform in New York City, 1894–1914

JOSEPH J. VARGA

MONTHLY REVIEW PRESS
New York

Copyright © 2013 by Joseph J. Varga
All Rights Reserved

Library of Congress Cataloging-in-Publication Data
Varga, Joseph J.
  Hell's Kitchen and the battle for urban space : class struggle and progressive reform in New York City 1894–1914 / Joseph J. Varga.
    pages cm
  Includes bibliographical references and index.
  ISBN 978-1-58367-349-2 (cloth : alk. paper) — ISBN 978-1-58367-348-5 (pbk. : alk. paper) 1. Hell's Kitchen (New York, N.Y.)—History. 2. New York (N.Y.)—History—1898–1951. 3. New York (N.Y.)—Politics and government—1898–1951. 4. City planning—Social aspects—New York (State)—New York. 5. City planning—Political aspects—New York (State)—New York. 6. Social structure—New York (State)—New York—History. 7. Social classes—New York (State)—New York—History. 8. Progressivism (United States politics)—History. I. Title.
  F128.68.H45V37 2013
  974.7'041—dc23
                            2013019260

Monthly Review Press
146 West 29th Street, Suite 6W
New York, New York 10001

www.monthlyreview.org

5 4 3 2 1

Contents

Acknowledgments 7
Introduction: A Death in the Kitchen 11

CHAPTER 1: SPACE AS HISTORY 19
Hell's Kitchen 19
Uneven Geographic Development 27
The Production of Space 31
Implications 37

CHAPTER 2: RESTRUCTURING PROGRESSIVES 45
Progressive Vision and Visibility 49
Visibility and Occlusion 61
Representational Space and Performance 74

CHAPTER 3: WHEN HELL FROZE OVER 87
Policing the Boundaries 103

CHAPTER 4: HOUSING AND VISIBLE SPACES 121
Imagined Spatial Communities 127
Kitchen Space 139

CHAPTER 5: SPATIAL ECONOMIES 165
Working in the City 169
The Laboring Body 180

CHAPTER 6: HELL, DEATH, AND URBAN POLITICS 203
    The Restructured Spatial Container 207
    Self-Perception and Citizenship 216

Conclusion: The Spatial Production of Desire 231
    The Future of Cities 235

Bibliography 239
Notes 247
Index 267

Acknowledgments

This project began as research with Diane Davis at the New School for Social Research's Graduate Faculty, sometime around 2003. Diane inspired me to think like a sociologist, and to dig into the fabric of the city. But it really started well before that, with my own interest in the working-class communities I grew up with, and the immigrants and workers in places like New Brunswick and South River, New Jersey, who chased their own version of the American Dream in space not of their own choosing. I was fascinated with how working folks used their collective power to build communities, defend their common interests, and fight for a better life. I learned in my work life the important battles that happen on the shop floor, when workers demand their rights and struggle for a better life. In that regard, this book is in part a tribute to all my coworkers in every job, but especially the crew at High Grade Beverage, Teamsters 701, who taught me lasting lessons in how workers control space, and made me laugh while doing it. Here's to you, Cuz, Richie, Carmen, and all the people who work every day to keep it all together.

In the six-plus years that I spent studying Hell's Kitchen, poring over documents, scouring resources, and walking its streets, I have leaned upon, and at times, annoyed many people. While I have always tried to keep my object at an objective distance, a mere mention of hell, kitchen, middle, west, or even Manhattan, would make me spin my head around. The kind staffs at the New York

City Department of Records library, the New York Public Library Branch at Eleventh Avenue and 48th Street, the Hartley House Settlement, and the McManus Democratic Club, among others, were always ready to indulge my often futile searches. Thanks to Brooklyn College's Edwin Burrows and Jocelyn Wills for pointing to archives I overlooked, and to the too-numerous-to-name army of adjunct faculty in the BC History Department who listened to my endless reiterations of the importance of spatial history, and for their conviviality. Thanks to Andrew Arato, Vickie Hattam, and José Casanova for trying to teach me to be rigorous in my application of social theory, though I'm not sure I ever fully absorbed the message. A special thanks goes to two people no longer with us, Chuck Tilly and Ari Zolberg, who for me were models of teaching, scholarship, and humanity. This project never would have come close to fruition without Oz Frankel, whose sense of humor, and sense of history, both inspired and constrained. Thanks also to my Indiana University colleagues Paul Mishler, Gerrie Casey, Thandabantu Iverson, Marquita Walker, and Bill Mello for reading and keeping me sharp, and Irene Quiero-Tajali, Lynn Duggan, Rae Sovereign, and Mark Crouch for general support and encouragement.

I am especially indebted to Michal Yates and the crew at Monthly Review, who had confidence in this project. Thanks to Martin Paddio and Scott Borchert for work on the manuscript, and to Erin Clermont, who really turned my sometimes aimless musings into workable prose.

Thanks to Peter Nekola and Laura Roush for being readers, critics, advisors, and, mostly, for being friends when the world seemed bleak. I love you both. And for my sanity, I must send a shout-out to Joe Kneafsy and Mark Beebe and good friends old and new in Brooklyn and Bloomington. And of course, thanks to my wonderful mom, Nancy Ann Varga, and to Al, Kathy, Scott, Barbara, and Marshall for their limitless love and patience. Much of the inspiration for this book, and for all the work I try to do with and for working folks, is inspired by my dad, Julius Varga. He came to this country at 17, after years as a Displaced Person, arriving with little education, very little knowledge of English, from a childhood ripped apart by war. He sought his own version of the American Dream working on a machine in a factory. I'm not sure if he ever realized his dream before his death in 1987, but he did raise three

Acknowledgments

boys, loved my mom, and did the best he could, which is all we can ask. I think he would be happy with this book, but probably think I was getting uppity, and knock me down a peg. This is for you, Dad.

I would thank my everything, my love, my Rebekah, too, but she always already knows.

INTRODUCTION

A Death in the Kitchen

ON APRIL 23, 2008, two men, David Daloia and James O'Hare, were cleared in a New York City court of all criminal charges in a case involving felony fraud, a dead body, and an obscure 1954 Department of Health law requiring burial or incineration following "a reasonable time" after death. In a case that delighted New York City's twin tabloids, the *New York Post* and the *Daily News*, Daloia and O'Hare had been arrested at a Pay-O-Matic check-cashing outlet for attempting to cash the Social Security check of their then-deceased friend, Virgillio Cintron. Claiming that they were unaware that Cintron had died, and having cashed checks for the incapacitated Parkinson's victim in the past, the pair, dubbed "Dumb and Dumber" by both newspapers, had dressed Cintron's corpse, signed his check, and wheeled his body in a desk chair from the Manhattan apartment Cintron shared with O'Hare on 52nd Street to the Pay-O-Matic at Ninth Avenue. The pair attracted curious onlookers, and caught the attention of a New York Police Department detective eating his lunch across from the check-cashing store. While both the *Post* and the *Daily News* played on the alleged "stupidity" of the crime, and the clear marginality of the defendants, the more staid *New York Times* highlighted the case as a throwback, a vestige of the old Hell's Kitchen in the rapidly restructuring Clinton District. Charges against the pair were dropped when a coroner could not determine a precise time of death, allowing both men to plead that they believed Cintron was alive on their trip. Virgillio Cintron was eventually buried through the intervention of a local funeral home, as neither he nor his family could afford interment. The check that O'Hare and Daloia attempted to cash was valued at $335.

Daloia, O'Hare, and Cintron could indeed be viewed, in 2008, as vestiges of a bygone era, ghostly images of the rough-and-tumble days of Hell's Kitchen, now commonly referred to as the Clinton Historic District, with sections designated grandly by realtors as "Clinton Heights." Yet the story is of interest for more than this. It connects directly to this book, which is an exploration of Manhattan's Middle West Side, or "Hell's Kitchen," between 1894 and 1914. The tone of the tabloid press, their emphasis on the "colorful" characters of Daloia and O'Hare, echoes techniques used by the turn-of-the-century press, and their accounts of the ethnic working class of that period. Such accounts continue to shape the perception of consumers of news tabloids, as they did before, during, and after the Progressive Era. But the image of the three men, poor white ethnics clearly struggling to make it through the day, even to the point of pulling, as the *Times* called it, a "Weekend at Bernie's" caper, amid the restructuring of their neighborhood, speaks to the point of this book directly. As the *Times* pointed out, these men and their struggle are, in fact, vestiges of a disappearing population on New York's Middle West Side. Their "type"—poor white renters, what used to be known as working class—slowly being pushed out of the area by those twin urban storms—development and gentrification. For some, this is the nature of the city; things change, and a world-class city like New York must be in the forefront of urban development. In the global economy, so we are told, there are winners and losers. For others, development and gentrification are catch-all phrases of demonization, the twin processes that force out authentic city residents, destroy historic districts, and de-democratize the neoliberal city of global homogenization. Whatever view one takes of the process, it is axiomatic that in the city, any city, different space is produced over time. For the Middle West Side, the physical space that provided affordable rental housing for people living on the margins of the city's economy, people like Cintron, O'Hare, and Daloia, is being replaced. But still they remain, as do businesses that serve them, like Pay-O-Matic.

Despite that things have changed, other physical spaces, produced between 1894 and 1914, are still there. DeWitt Clinton Park, built in 1901, remains, though in an adulterated form, as do many of the pre-1901 tenements, whose construction form was outlawed. Post-1901 tenements still stand among the high-rise condo developments, but at rental prices far higher, relatively, than they originally commanded. But the urban poor, defined and classified by progressive reformers, remain, though in much reduced numbers. Even some businesses that started in the nineteenth century, like United City Ice Company, still serve the local population, though United now provides bagged ice produced by machines and delivered to local stores, rather than the cut-up chunks of Hudson River ice it used to distribute to local residents. What remains also is the human use of physical spaces—the way that material, altered by natural and human activity, is produced into the spaces of people's everyday lives. The produced space also changes. DeWitt Clinton Park, a triumph of urban reformers in 1905, is now one of many underfunded, underserviced city parks, managed in part by local activists. The space has changed, and so has what it means. The combination of the physical space, its perception and how it is used and understood, is what makes it produced space, space as a social product, and space as history.

French sociologist and philosopher Henri Lefebvre, in his complex work *The Production of Space,* claims: "If space is produced, if there is a productive process, then we are dealing with *history.*"[1] This history, for Lefebvre, is not the "causal chain of historical events" or a structured, ideological sequence, but is rather the constant result of the combining, recombining, and decomposition of productive forces. These forces include physical material, labor, knowledge, and technique. Thus the history we are seeking here is not simply the history of human social organization, the workings of economic, social, and political power structures. And, of course, there is no one singular history of New York City, or the Middle West Side, from 1894 to 1914, or for any other demarcated temporal span. Nor are there many histories, endlessly multiplying narratives based on perspective, focus, and sometimes ideology. There is only history, the composition and decomposition of knowable forms. These forms are knowable through archival information, the knowledge base of the historian. However, the main thrust of this project is drawn from a different archive than most previous histories, not a newly discovered box of correspondence, or dusty, long-unread government documents, but the real physical space of the Middle West Side, history as a social product.

The book begins by introducing the reader to the Middle West Side and its working-class population during the Progressive Era. In the initial chapter, work's relationship to existing urban spatial theory is charted. Introducing certain key concepts, such as the production of space and spatial citizenship, the chapter describes the relationship between space and evaluation, between spatial restructuring and political power, and between the physical environments and urban populations From the start, the book presents physical space as an agent, a co-actant in the process of spatial production and the production of history. This first chapter leads into the subsequent discussions of particular spaces and particular political formations that follow.

Chapter 2 takes an overview of the U.S. Progressive Era and its historiography. Here I suggest that a genuine analysis of spatial practice and its consequences can deepen our understanding of this important period in American history. A review of historical writing on the Progressive Era reveals how spatial practice has been either ignored or relegated to issues of place or setting. Though recent scholarship on the era attempts to place the work of progressives within a global or at least Atlantic context, few works examine either the processes of the production of space or the spatial redeployment of progressive actions by urban residents. Chapter 2 also deals with the historical context of spatial restructuring on the Middle West Side through a brief review of international, national, and local political economy. It concludes by focusing on New York City, and the variety of factors that influenced both the history of the Middle West Side and its subsequent restructuring, placing the changing built environment and changing conceptions of citizenship within the larger framework of reformism, immigration, city economics and politics, and the expanding culture of consumerism and entertainment.

Chapter 3 charts the spatial production of difference, showing how uneven geographic development and the production of space contribute to the physical and mental construction of neighborhoods, paths, nodes, and "frozen zones." Using several instances of

Introduction: A Death in the Kitchen 15

ethnic and racial confrontation in combination with the activities and perceptions of city agents, particularly the New York City Police Department, I show how the spatial practice of the commercial city combines with outside conceptions and internal representations to produce difference. The production of difference, in this sense, also produces the language of understanding employed by outsiders and authorities in building their perceptual understanding of areas like Hell's Kitchen.

Chapter 4 deals mainly with the efforts of social reformers to impose their vision of "proper citizenship" on urban residents through the reconstruction of space. Here the emphasis is on the conceived spatial constructions of reform and the "gospel of moral environmentalism." It was believed that the construction of the proper environment would produce the proper citizen, one fully equipped to deal with the expanding consumer economy and an increasingly complex social system. Though many reformers of the era focused their attention on the "new immigrant" populations of northeastern cities, others turned their attention to areas of entrenched and durable poverty and inequality, utilizing the most contemporary methods of analysis to attempt to create a new vision of an integrated society where citizens would be active participants in the political process.

Chapter 5 takes up the issue of economics and its relation to spatial form. Focusing on the docks, factories, and small enterprises in the area, it shows how the spatial practices of accumulation create a language of insecurity that governs the self-conceptions of Middle West Side residents. Testimony of area residents, and indeed corresponding reports from other industrial cities, consistently rank economic insecurity, the fear of losing one's job and place of residence, as the primary concern among the wage-earning population. In this chapter, I demonstrate how an already globalized economy promoted insecurity in a period when the transition to a functioning welfare state was in its infancy. The economic insecurity of Middle West Side residents caused many to seek security in small-scale property ownership, often at the cost of short-term comfort.

In chapter 6, several incidents demonstrate the political agency of Middle West Side residents, and how this agency is shaped by spatial practice. The incidents considered depict an unwillingness of Middle West Side residents to accept either the condescension of philanthropic reformers or the outsider perception of their neighborhood as a zone of vice and immorality. The chapter explores the spatialized nature of emerging subjectivities and the shifting definitions of class, ethnicity, and gender that are generated by spatial restructuring. Spatial restructuring and the relational construction of place and locality serve to create the performative spaces in which Middle West Side residents utilize their ethnic, gender, and class positions to reinterpret the normative framework of Progressive Era citizenship.

In the final chapter, I suggest effects, both short and long term, of the constituent spatial restructuring experienced by Middle West Side residents. In this view, spatial production creates not the spaces of identity or redeployment suggested by many urban historians and sociologists, but instead generates specifically urban, modern forms of knowledge based within the language of uneven geographic development. Urban space is history, and history is the urban space itself.

It has been twenty years since critical geographer Edward Soja, in his landmark work *Postmodern Geographies*, urged social theorists to "take space seriously." Soja built on the work of Lefebvre and others to promote the inclusion of space as an "ontological category of understanding" and as something that is both produced and produces. Though many urbanists, historians, and social theorists have heeded the call to take space seriously, most have produced work that describes, explains, or analyzes the spatial formations of urban development as either the process of capitalist accumulation or the way space is used and re-territorialized by urban inhabitants to create community, identity, or some combination of the two. This spatial history examines how space is produced during a specific period and under specific conditions, and tries to capture the affective nature of forces of production that have been largely excluded in other histories. In other words, it really takes space seriously, and examines the way history is produced by the relationship between the physical built environment, the natural landscape, and human actors. In so doing, the project tackles several important questions relevant to both urban history and urban sociology. First, how is space produced, as both a product of social relationships and as the result of the interaction of human and non-human? Second, how is space understood, how is it perceived, conceived, lived and altered?

Introduction: A Death in the Kitchen

Third, how does spatial production affect the way human actors judge and evaluate urban space and its development, and what effect does this have on such important categories as democracy and citizenship? And finally, how does taking real space seriously inform our view of crucial urban questions, and perhaps re-form our use of key analytic categories such as community, place, neighborhood, and scale? In taking space seriously, I hope in this book to encourage other researchers to utilize the production of space as a category for understanding and analyzing other complex processes, such as the formation of class solidarities and fragmentations, and the shaping and reshaping of urban communities. Serious indeed.

CHAPTER ONE

Space as History

Both collective memory and collective identity are the effects of inter-subjective practices of signification, neither given nor fixed but constantly re-created within the framework of marginally contestable rules for discourse.
—JONATHAN BOYARIN

People without wants are poor.
—HONORÉ DE BALZAC

HELL'S KITCHEN

IN HIS HISTORICAL SKETCH of New York City's Middle West Side, written in 1912, social worker Otho Cartwright describes not only the character of the district, but, as important, how the district was perceived by outsiders:

> The district of which we write has been known for many years as the scene of disorders, of disregard of property rights and public peace. Certain it is that in the minds of New Yorkers who live outside the district . . . as well as in the minds of the police authorities, there still lingers a tendency and doubtless a liking to think and speak of the district by the nickname that disorders, rioting, and crime won for it in the early days of its settlement, namely "Hell's Kitchen."[2]

Referring to the residents of the area and their perceived apathy, Cartwright grimly quotes the French writer, Balzac: "People without wants are poor." From his perspective as a social worker at the Bureau of Social Research, Cartwright surveyed a district of "disheartening inertia," "social neglect," and a population that had, in his words, "accepted the conditions of their environment."[3]

As a foot soldier at the cutting edge of the Progressive movement, Cartwright, like many of his contemporaries, tended to emphasize the worst aspects of his objects of study, and to analyze them in terms of physical environment.[4] Working for reform organizations that were, in large part, funded by private interests, these Progressives, armed with clipboards, surveys, and the latest social theory, set out to catalogue the quantitative and qualitative conditions of the urban poor. Their work, while problematic, has left a valuable archive by which to evaluate the history of urban America during the Progressive Era. As Cartwright and other social workers demonstrated, the Middle West Side was certainly a district with more than its share of urban problems. But by 1912 it was also, like many other areas of Manhattan, part of the great experimental laboratory of the Progressive movement. Participating in what the urban historian Stanley Schultz has dubbed "the gospel of moral environmentalism,"[5] Progressives like Cartwright used their statistical and narrative evidence to advocate for changes in the built environment in areas like Hell's Kitchen. For these Progressives, providing the poor with the proper living environment was seen as the surefire cure for the ills of urban communities whose development had resulted from the demands of the private market. Attempts to ameliorate these conditions took a variety of forms. From the creation, in 1884, of the New York City Tenement House Commission, to the formation of the West Side Improvement Association in 1907, Hell's Kitchen[6] had been subject to spatial restructuring projects brought about, in part, through the efforts of Progressive reformers. The passage of tenement laws, new park and bathhouse construction, dock repairs, the locating of settlement houses, and street cleaning were just some of the projects carried out between 1894 and 1914 to provide a proper environment for the residents of the Middle West Side. It was hoped that, as social worker Katherine Anthony explained, "the appearance of respectability would create the desire for respectability" among the "less ambitious" West Side residents.[7]

Space as History 21

The spatial archive: a tenement with business on the ground floor. Note Ben-Hur Stables in background. (Milstein Division of United States History, The New York Public Library, Astor, Lenox, and Tilden Foundations)

Cartwright's "people without wants" occupied the area from Seventh Avenue to the Hudson River,[8] bounded on the south by Thirty-fourth Street and on the north, according to most sources, by 54th Street.[9] Known from the Dutch period as the Great Kill District,[10] it was a mixture of hilly forest, swampland, and streams, the largest of which gave the district its early moniker. The undesirability of much of the landscape for residential development, as well as the district's proximity to the Hudson River waterfront, has linked its developmental history to industry, shipping, and manufacturing, which greatly contributed to its reputation as "dingy and noxious."[11] By the late nineteenth century, the area was dominated by low-lying factories, warehouses, stables, and piers west of Tenth Avenue, and by tenement apartments of the "dumbbell" or "railroad" type east to Broadway. Indeed, as Cartwright states, "From an architectural standpoint the district has neither salient features nor real uniformity."[12] Though Cartwright places much of the blame for this state of affairs on the "lack of a proper building plan," city planning itself is partly responsible for the perceived lack of character. As urban historians Hendrik Hartog, Jon Teaford, and others have demonstrated, New York City was part of the municipal revolution of the nineteenth century, as the city government took an increasing role in providing street plans, construction, transportation infrastructure, and other services, in combination with the state of New York.[13] The monotony and lack of "quaintness or color" that Cartwright bemoans were as much the result of New York's grid plan of 1811 and the uneven level of street repair, coupled with the provisioning of elevated railways meant to spur development and population, as they are the result of the dominance of private interests and the chaotic development of unregulated capitalism.

Far from being the result of a lack of planning, the spatial environment of the Middle West Side was the result of competing interests with differing views of the value of space. Proximity to the waterfront made the area prime space for certain industries. Tenement landlords and the system of subleasing produced an inadequate and often unsafe housing stock.[14] City and state government directed resources to the creation of the parks that framed the area (Central and Riverside), while neglecting to provision the less politically influential residents of the Middle West Side with open space and the real estate value it creates in its wake. Reformers utilized statistical studies and the latest in social theory to demand the construction of playgrounds and bathhouses, as well as rules governing the form of tenements. And of course, the residents and merchants of the area turned the space of the Middle West Side into the places of their daily rounds. Thus while one must pay strict attention to the dominant mode of spatial practice, the demands of a capitalist economy, critical geography also realizes that space at the level of the urban scale is, in the words of Edward Soja, "a multilayered geography of socially created and differentiated nodal regions nesting at many different scales around the mobile spaces of the human body and the more fixed communal locales of human settlements."[15] Though the actual built environment is the result of spatial practices of power and domination, the production of space is something quite different. Thus while it remains true that the records of the deeds of Progressives form a valuable historical archive of the period, the built environment itself, the space of daily life, may indeed tell us more about both the reformers and the population they intended to reform.

The occupants of the tenements of the Middle West Side at the turn of the century were, unlike their more colorful and well-studied new immigrant neighbors of the Lower East Side, nearly evenly split between foreign born and those born in the United States. They were predominantly "white," and predominantly descended from parents of Irish and German origin. Many of the Irish families had arrived as part of the famine migrations of the 1850s and 1879, and many German families descended from those arriving during the great wave of German immigration between 1817 and 1853. Many of these families had lived on the Middle West Side for several generations by 1900, and were joined by new Irish and German arrivals seeking work on the waterfront. These two ethnic groups dominated the local population, but by the turn of the century they were joined by smaller clusters of new immigrants and shared space with, among others, small pockets of Jews, African Americans, and Anglo-Saxon "natives." As with most urban neighborhoods, residents were drawn to their own ethnic and racial cohorts, with clusters centered in church parishes, divisions that often resulted in open violence, but there were also numerous examples of interethnic cooperation on the Middle West Side. Residents shared the common experiences of economic insecurity, lack of social services, and being commonly perceived by outsiders as dangerous, and the area was widely known by the city's middle and upper classes as one to avoid. Perhaps the most commonly shared experience was the lack of respect Middle West Side residents received from city authorities, a phenomenon that had strong, if rare and brief, unifying effects.

Far from being a population without wants, residents of the Middle West Side make up a fractious but coherent community whose individual and collective wants were structured and contained by the physical environment of their neighborhoods and their own place as a community within the processes of uneven geographic development that shaped the built environment. Operating within what geographer Richard Peete terms the "daily prism" of segregated locational scales, Middle West Side residents formed their wants within a specific framework of spatially determined action. Those actions were framed by the historically contingent spatial conditions of the area, where, by 1900, 230,000 of the 270,000 residents lived in buildings officially classified as tenements. Further, residents made their daily rounds in an area where a lack of regulation had led to mixed-use development, with tenements often abutting factories and waste drainage sites for the runoff from slaughterhouses.[16] As mentioned, Hell's Kitchen had garnered a reputation as both slum and red-light district, leading to negative conceptions of residents by outsiders and city officials. As a result, residents expressed a deep dissatisfaction with city services, and a deep distrust of city officials, particularly police officers. These officials returned the sentiment, often viewing residents of "the Kitchen" as criminals. As sociologists John Logan and Harvey Molotch point out, "Location establishes a special collective interest among individuals."[17]

As well, Progressive reformers who encouraged city residents to demand efficient service from their municipal government shaped wants in the form of political demands. Many reformers saw the relationship between built environment and human behavior as direct and unambiguous.[18] Trained by men like the social theorist Simon Patten, they were deeply influenced by European ideas regarding city planning and proper physical environment. Progressive social workers, particularly after 1900, turned increasingly from focusing on the morals of the urban poor to an emphasis on the scientific basis of poverty. Seeing the built environment as a key to combating entrenched poverty, many Progressives not only advocated for improved physical conditions, but also sought to form a public among the urban poor that would demand improvements of city authorities. Seizing the opportunity provided by the perception among the public that big business needed to be regulated and that party politics was deeply corrupted, these reformers used their positions as experts to promote spatial restructuring and encourage urban citizens to participate in the daily running of the city. They believed that even the urban poor, provided with the proper tools and space, could become "rational citizens," and form what New York Bureau of Municipal Research founding member Frederick Cleveland dubbed "a socialism of intelligence."[19]

For the residents of the Middle West Side, wants were further contained and promoted by the structural conditions at the level of

Steam engine on Eleventh Avenue, running through one of the most crowded neighborhoods in the city. (Milstein Division of United States History, The New York Public Library, Astor, Lenox, and Tilden Foundations)

local, city, state, and federal government, which all contributed to the spatial restructuring of the district. The period under investigation, 1894 to 1914, is one in which the government at all levels steadily increased its activities in the areas of social provision, infrastructure improvement, and the regulation of immigration.[20] Government at all levels increased in its ability to regulate the private marketplace, as seen in the battles with developers over housing, utilities, and transportation, and in larger battles over monopoly control of economic sectors. The period also saw the rise of reform at the local and state level, as politicians increasingly forged partnerships with nongovernmental Progressive groups, bearing the mantle of reformers whether they were connected to machines or not.[21] The attacks on patronage and on the political party structure, part of the attempt to create a rational public sphere, produced new political formations, such as fusion candidates and independent commissions whose goals were to break the power of localized party political organizations. These changing political circumstances greatly affected the process of spatial restructuring and the attitudes of residents toward government at all levels.

What we are interested in, then, is directly concerned with measuring the effects of such spatial restructuring on the residents of

The Ninth Avenue elevated train ran through Hell's Kitchen. Until 1902, steam trains ran on the line, dropping hot ash on pedestrians below. (George Grantham Bain Collection, Library of Congress)

Manhattan's Middle West Side in the period between 1894 and 1914. Spatial restructuring is defined here as changes in the built environment brought about through the combination of private market interests, government agencies, and reform organizations. The main question we address is, how does the restructuring of physical space affect human activity? We attempt to answer the question through an examination of the processes of uneven geographic development and the production of space.[22] We aim, first, to achieve a deeper understanding of how the physical built environment structures urban life, and second, to find an adequate way to understand the particular spatial history of the period under study. In the case of the former, there is little dispute as to the importance of the built environment in urban history. But what is less accepted is how space itself acts in the process of structuring urban political alliances. As to the latter question, linking changes in the built environment to the concept of the production of space opens new opportunities not only for understanding the history of urban development during the Progressive Era, but can be usefully applied to contemporary issues, such as disputes over school district boundaries and localized secession movements.

Making the connection between space and urban politics requires not only examining how space is produced, but also the "difficult prospect" that such a process of production is, as anthropologist and Marxist geographer David Harvey suggests, "constitutive of the very standards of social justice used to evaluate and modify" its use and subsequent alteration.[23] In other words, how does the spatial environment produce the very discourse that is used to guide, direct, and control spatial restructuring and what effect does this discourse have on residents? This suggests that the very process of spatial restructuring and the decisions that guide such a process are framed within a language that is produced by perceptions and conceptions of the space itself. When city planners and government officials make decisions about where to invest in urban restructuring, they are influenced by their preconceived ideas about the physical space and its inhabitants. Coded descriptive language, such as dangerous, pleasant, blighted, up-and-coming, struggling, bedroom community, concern not just the area but is seen as descriptive of residents as well, serving to produce perceptions that then govern decisions of future investment. Aside from affecting investment decisions, these codes also play a role in the formation of the self-conception of residents, particularly in their relationships with authority and with notions of citizenship. In this book we will be examining the contradictory ways that technologies of control deployed by reformers and government attempt to shape a temporal horizon for residents that is then weighed down and reinterpreted within the spatial order. In order to begin answering this difficult question concerning the relationship between space and citizenship, it is necessary to turn to theories of spatial ontology to uncover the role of spatial restructuring in the formation of conceptions of citizenship.

UNEVEN GEOGRAPHIC DEVELOPMENT

"UNEVEN GEOGRAPHIC DEVELOPMENT is a concept deserving of the closest elaboration and attention," states David Harvey.[24] For Harvey, uneven geographic development concerns the "production of spatial scale" and the "production of geographical difference,"[25] with both components interacting to produce space and reproduce social relations. But as Harvey points out, "relevant scales" are never produced outside of the natural components or influences that structure all human geographic alteration. Scale here refers to the human understanding of space and terrain, how we order things in space at different levels, such as neighborhood, city, region, and state, to name just a few. Our understanding of the scalar processes of the production of urban space must first include an analysis of the relation of the natural contours and terrain to the construction of the built environment. For New York City's Middle West Side, proximity to the Hudson River, as well as the original swampy conditions of the area made it a less than ideal location for upscale residential development, and subsequently the area was utilized mainly as a port, site of small-scale production, and the location of low-income, high-density housing, something we deal with more extensively below.

From an understanding of the natural terrain, our understanding of the production of space then moves to an examination of scalar spatial production related to the human "pursuit of goals and organization of collective behaviors." Here, such endeavors as the production of goods and services to meet human needs are carried out at different scales. "Households, communities, and nations are obvious examples of contemporary organizational forms that exist at different scales."[26] Thus to properly understand the processes of uneven development and the production of space, it is vital to approach the problem through an examination of the differing levels of scalar production within the urban spatial environment. Indeed, Harvey suggests that we can understand urbanization itself as " a manifestation of uneven geographical development at a certain scale."[27] Human beings, according to this approach, produce geographic scales by altering nature to meet their needs, wants, and desires, and for organizing collective behaviors. Geographic scale, and for our purposes, urban scale, is organized at different levels to meet different needs. As geographer Eric Swyngedouw states, "Spatial scales are never fixed, but are perpetually redefined, contested, and restructured in terms of their extent, content, relative importance and interrelations."[28] The long, slow process that produced the district known as Hell's Kitchen, with its polluted waterfront, aging housing stock, working-class population, and bad reputation, is exemplary of this process of scalar production.

One useful approach to the question of scale is Logan and Molotch's division of use and exchange value when examining the way spatial restructuring is carried out. They suggest that the differing visions of land use utilized by entrepreneurs and residents will clash, producing a struggle between those "planning and organizing themselves to make money" through the increase of aggregate rent levels, and those striving for "affection, community, and sheer physical survival."[29] This struggle produces the microscales of area development, as residents push for a livable environment against the desires of developers to turn a profit. These interests can coincide, often when governmental or professional authorities intervene. For example, Middle West Side residents' desire for usable spaces of recreation often clashed with not only the larger scale of urban transport and the needs of industrial production, but also with the conceived scale of the functioning, rational city promoted by urban reformers. Often when debate occurred over the provision of recreational space as a use value for residents, these needs were trumped by the exchange value of the space for economic production. The case studies presented in subsequent chapters chart the increasing instances of such intervention, mediating changes at the scale of citywide transportation and at the neighborhood scale of providing a livable environment.

For the Middle West Side, uneven geographic development is the product of the variety of influences that contribute to the construction of the built environment at the local or neighborhood level. As Harvey and others have shown, all built environments are the result of uneven allocation of resources, where the various factors of production are distributed based upon power and struggle. For uneven development at the urban scale within a larger system of production, distribution, and consumption organized under the principles of capitalist investment, the built environment will reflect the needs of the dominant mode of production. Development thus is never the result of unseen market forces, nor is it the result of natural processes of allocation described and theorized by the Chicago School of urban sociology, but rather is the result of a complex interaction of interests dominated by those collective actors whose maintenance is dependent upon the reproduction of existing social relations. As

Harvey states, "Transformations of space, place and environment are neither neutral nor innocent with respect to practices of domination and control. Indeed, they are fundamental framing decisions" that largely determine the structural possibilities of human aggregate populations.[30] But, as even Harvey concedes, this process is "replete with multiple possibilities."[31] Thus to understand uneven geographic development in the Middle West Side between 1894 and 1914, we need not only to understand the "dismal logic of the consequence of capital accumulation" described by Harvey, but the more complex logic of the interactions and intersections of competing interests, and of multiple, often incommensurate, factors that are unplanned and often misrecognized.[32] Thus, Harvey's approach to uneven geographic development is inadequate for a synthetic understanding of the process of spatial restructuring, but his concept of the generative or constitutive nature of space is vital to understanding space as both product and producer.

To explain what is meant by the generative or constituent nature of space, it is necessary to think of space not as the primary location of human activity but as a primary ontological condition of human understanding. In this understanding, space is more than physical context in location. This conception of space takes our perception of location, what philosopher Martin Buber termed "our primary setting at a distance," as, in Edward Soja's formulation, the "original existential capacity to separate the individuated Human Being from the whole of nature."[33] Here, distance, proximity, location, and their perception make up the primary locus of human understanding—in other words, how we locate ourselves as beings-in-the-world, and how we formulate abstract ideas and conceptions regarding our relationships to that world. Space, then, is understood as a primary locus of consciousness, producing both our understanding of our relationship to the physical world as well as producing the languages we utilize to understand and alter that world.[34] Therefore, changes in the spatial environment, as Harvey suggests, will govern or constitute the abstract categories we construct, such as justice or legality, or indeed citizenship, and alter our perception of ourselves as the environment itself is restructured. In this view, space is neither separate from the temporal, nor an independent category, but the means by which the spatiotemporal matrix, how we orient ourselves in the world, is constructed in conjunction with temporality.

THE PRODUCTION OF SPACE

A FULL UNDERSTANDING of uneven geographic development requires an examination of the production of space as a primary factor in the formation of collective understandings. This means we must not only investigate the activities of private market interests and government, but how these activities create meaning, how they combine in a variety of ways to produce not just the physicality of the built environment, but the complex set of signs and symbols that allow navigation and deployment of this environment at differing scales. Scale here refers not only to the physical measurements of household, community, and city, but to the layers of memory and interpretation taking place at different levels of collective understandings.[35] So scale can mean ways of measuring and delineating districts, regions, and neighborhoods, but it also includes how those different physical demarcations have varied meanings for different users of the space, based on their histories. For an area like Hell's Kitchen, the disparate languages of description of the region deployed by social workers, residents, police officers, journalists, and politicians illuminate the different levels of understanding at work in various public spheres.[36] As urban planning professor Robert Beauregard states in *Voices of Decline*, the discourses that define urban areas function as ideological markers that "shape our attention, provide reasons for actions, and provide a comprehensive, compelling, and consistent story" regarding regions of difference.[37] Following Henri Lefebvre, we can approach the production of space and its ability to generate or constitute the discourse that governs its restructuring by viewing space on three levels: lived, conceived, and perceived. This method will demonstrate the ways in which space as built environment acts as both producer and product, as produced space penetrates the consciousness of urban residents and shapes the way they devise knowledge, construct languages, and frame identities.

In his work *The Production of Space*, Lefebvre emphatically states, "If space is produced, if there is a productive process, then we are dealing with *history*."[38] This history, for Lefebvre, is neither a sequence of causal chains nor teleology, but an ever-changing dialectic of nature and productive forces. It is within this essential relationship that space is produced. Lefebvre draws a distinction between the "problematic of space" and "spatial practice," where the emergence of specific, historical scales of urban space are produced at the intersection of the two. "The problematic of space is comprised of questions about mental and social space, about their interconnections, about their links with nature on the one hand and with 'pure' forms on the other."[39] These questions, according to Lefebvre, comprise the conceived space of his eventual triad of production, the representations of space that comprise the realms of planning, abstract knowledge, engineering, and other activities that frame spatial restructuring. Indeed, Lefebvre claims that representations of space are the "dominant space in any society," the locus of social power. Representations of space comprise that which is conceived within the context of accumulated and layered social knowledge. These representations, how space is imagined before it comes into physical being, have a very practical impact on design, planning, and restructuring. According to Lefebvre, conceptual spaces, informed by effective knowledge and by ideology, play a substantial and specific role in the production of space. In Lefebvre's telling use of the term *architecture*, it is not a specific structure or building, but "a project embedded in a spatial context (lived space or spatial practice) and a texture which calls for representations that will not vanish into the symbolic or imaginary realm." Representations of space are particularly important in describing changes in the built environment. They are, as Lefebvre points out, "shot through with knowledge," with a conceptual logic of the prevailing conditions of knowledge and power, and thus objective as well as ideological. It is this objectivity, ideologically informed, that drives the process of restructuring by reconceptualizing space according to shifts and changes in the forms of shared knowledge. The Progressive Era's emphasis on the link between space and morality is highly indicative of this process. The driving idea of the urban planner, that one can change behavior by changing space, is the very epitome of the power of representations of space.

Spatial practice, however, consists of the use of spatially created codes to form a cohesive yet not necessarily coherent set of signs and symbols that enable the daily practice of meeting needs and reproducing social relations. This "lived space," the daily reality or routine, comprises what the urban theorist Kevin Lynch described as the paths, nodes, edges, districts, and landmarks that orient movement and connect networks on the urban scale.[40] It is not, for Lefebvre, how space is "lived," the daily experience of the urban subject, but urban space as a form of living, specific to the needs of the dominant mode of production. Spatial practice is defined less as the everyday use of the city, and more as the built environment, the result of various factors through which the city is negotiated. These factors include, but are not limited to, the actions and manipulations of real estate speculators, urban planners, private developers, and advocacy groups. Most important, spatial practice, or lived space, is the primary location of the spatial code of historically contingent systems of modes of production, distribution, and consumption. "The spatial code is not simply a means of reading or interpreting space: rather it is a means of living in that space, of understanding it, and of producing it. As such, it brings together verbal signs (words and sentences, along with the meaning invested in them by the signifying process) and nonverbal signs (music, sounds. evocations, architectural constructions)."[41]

Left there, Lefebvre's spatial theory would seem little different from the dismal logic of capitalist production examined by Harvey and others, where the needs of accumulation directly influence the conceptions of urban planners, producing spaces of flows that guarantee the aggregate levels of rent and accumulation that reproduce existing class relations. But Lefebvre's conception of the production of space adds a third analytic category, representational space, which opens his theoretical approach to different possibilities in the study of urban restructuring. Here, conceived and lived space are linked and disassociated simultaneously by representational space, space as directly lived not in the daily routine but "through its associated images and symbols." Representational space can be understood as the level of perception where subjective consciousness encounters the built environment and is penetrated and permeated by the myriad myths, symbols, signs, and forms of understanding that are produced by this very relationship. Representational space is defined as the everyday perception of the city as experienced by its users and allows for the creative reconceptualizing of the language formed to negotiate lived space and the abstract ideas that govern conceived space. Though not necessarily a liberating or emancipating concept, representational space is the location of possibility within a system of production dominated by needs of a spatial order whose codes are produced under specific forms of social and class relations. Thus in the triadic spatial interplay, space is lived, conceived, perceived, represented, negotiated, altered, judged, restructured, and ultimately experienced as a mode of knowledge construction and understanding as it is produced by all these factors.

Although Lefebvre describes representational space as the "dominated" or "passively experienced" space of the directly lived and perceived, I believe his formulation can be extended to view representational space as the active locus of spatial production. To understand this process, it is necessary to consider the body and its relation to the ecological system and look at the triadic levels of spatial production as a fluid process of interaction where representational spaces serve as the space that can be informed, but never captured, by ideology. Representational space refers to the almost limitless ways the world as perceived can be encountered through the filter of accumulated practices (lived and conceived) that work on the body to produce narratives of understanding. To explain this, Lefebvre uses an interesting metaphor, the heart: "The heart as lived is strangely different from the heart as thought or perceived." In other words, the heart functions as the organ that keeps us moving (spatial practice) and is conceived, under current forms of knowledge, in scientific terms, as a potential site of medical practice. But the perception of the heart is bound up with its biological function, its physical existence, experience of its workings, and the metaphorical conceptions (love, bravery) that govern the language used to communicate regarding it. Representational spaces are this realm of perception, the bodily encounter with the world of objects. It is the location where cultural production interacts with bodily experience to produce spaces that cannot be contained by either the logic of the conceptual or the orientating influence of spatial practice. As Lefebvre put it: "Representational spaces need obey no rules of consistency or cohesiveness. Redolent with imaginary and symbolic elements, they have their source in history—in the history of a people as well as in the history of each individual belonging to that people." Of course, conditioned by bodily experience and cultural symbols, representational space is as likely to reflect and justify ideology as it is to counteract it. Yet in the triadic formulation it seems to contain the kernel of possibility, the avenue to analysis of that which potentially escapes the coherent and cohesive.

At the level of practical application in the study of urban processes, Lefebvre's spatial theory opens up different ways of understanding uneven geographic development. It helps in explicating Harvey's difficult proposition that the language utilized to "modify and evaluate" urban processes is constituted by the very practices and space produced by uneven geographic development. Writing in France in the late 1970s, Lefebvre suggests that modern spatial practice might be defined as "the daily life of a tenant of a government-subsidized housing project."[42] For the purposes of this study, we can substitute the Middle West Side tenement dweller in 1900. The spatial practice of this period would determine the daily routine, the networks of connective nodal points that linked worker to place of employment, shopper to place of purchase, police officer to precinct, neighborhood to district, local place of production to national network of distribution, and all the various nodes, pathways, borders, and landmarks that made up the built environment. The development of waterfront docks, in various states of repair, along with the "shape-up" nature of the labor force of dockworkers, generated a language concerning this type of work and the waterfront itself. The Eleventh Avenue slaughterhouses and the railroad that ran down the center of Eleventh Avenue generated such language as "Abattoir Place," "Avenue of Death," and "Hell's Kitchen" itself. This language, generated by the physical surroundings, contributes to the conceived space of reform proclaimed in the "gospel of moral environmentalism." If, as Lefebvre claims, the Renaissance city code was embedded in the vanishing point and vanishing lines of that mode of production's "logic of visualization," then the Progressive Era city's logic of visualization, its conception of space, is driven by the ideological linkage of the space of the slum and the body of the lower, dangerous classes. It is these conceptualizations that produce the physical space of the settlement house, parks, and model tenements, and generate the discourse of proper, rational citizenship inscribed on the bodies of the urban poor.

Yet, contrary to such investigations as M. Christine Boyer's *Dreaming the Rational City*, the work of urban planners and reformers never produces urban space as a rational text, readable in a coherent and cohesive form. Though Boyer is rightfully critical of modernist forms of urban planning that emerge in the mid to late nineteenth century, her analysis of the planning mentality that produces what she terms the characteristic features of the modern city, "its alienating abstractions, rational efficiency, fragmented and malign configurations, ruptured tradition and memory,"[43] falls short of understanding the complexities of the production of urban space. Boyer acknowledges the failure of progressive urban planners to create the urban utopia many of them sought, but her critique, like that of Jane Jacobs, falls back on the assumption of a real or genuine space that is effaced by modes of power and domination. What Boyer and others fail to see is that forms of governmentality in the production of space are not the result of top-down efforts at control and domination. Rather, governmental control through the restructuring of urban space relies upon an already existing set of customs, practices, and histories among the population. Thus, as we will see in Hell's Kitchen, attempts at control can have unintended consequences when the conceptual visions of urban planners clash with the needs of urban residents.

This clash occurs at the level of representational space, the embodied, perceived space with the palimpsest of cultural production and history. This tension produces the liminal spaces that fall outside of spatial practice and conception, altered in their actual, day-to-day embodied usages, with unforeseen and unexpected consequences. Thus a model apartment on Ninth Avenue, built in 1880 to house tenement families, becomes, by 1900, a haven for "bohemian" women. Likewise, a settlement house at 38th Street and Eleventh Avenue, abandoned for lack of use, is seized by neighborhood residents for their own purposes at the same time local youths demand that the press cease in their use of the moniker "Hell's Kitchen." In the longer term, the spatially generated language that frames the area for both residents and outsiders produces a representational space where local residents demand their political recognition, and attempt to utilize space to achieve this ambition, by either owning in the area or moving out. In either case, their democratic impulses, the desire to participate as full citizens, is contained by the discourse produced by the built environment. What emerges is neither the closed system of ideological construction, in this case, the rational city and rational citizen, nor a liberating space of transgression, but a space of governmentality. Here, Foucault's "strange discourse" of madness is reflected in the discourse of reform.[44] The technologies of domination, embodied in lived and conceived space, interact with the spatially generated technologies of the self, or representational space, to produce internalized body habits and externalized knowledge that serve to structure democratic action.

IMPLICATIONS

IN HIS INSIGHTFUL WORK *City Trenches,* political scientist and historian Ira Katznelson argues that the separation of workplace and community politics into separate spatial spheres contained, in essence, the democratic impulses of urban workers and their families. Although American workers were as militant as their European counterparts when it came to union and workplace politics, ethnic and racial divisions at the level of community often served to limit the political activity of urban residents. According to Katznelson, this division meant that urban politics by the late nineteenth century was "a politics of competition between ethnic-territorial communities from which capital was absent."[45] The result of this competition, Katznelson writes, is that "by constricting the politics of class to the workplace, the urban system made challenges to the larger social order very difficult indeed." In describing the formation of "city trenches that protected capital," Katznelson performs a valuable spatial analysis of the American city from the mid-nineteenth century, as he uses the development of northern Manhattan to illustrate the way urban space frames the ability of residents to begin and carry out political sequences. Here, political sequences refers to the chains of activity, often based on grievances, related to demands for either recognition of rights and belonging or redistribution of wealth. Katznelson's approach to urban spatial practice informs his work, and provides valuable insights into the workings of politics, economics, and culture in the formation of city political coalitions. The objective separation of work and home in American cities produced a unique combination of class and ethnic politics in their respective realms, and historical evidence seems to bear out that the constant influx from 1830 to the present of different ethnic groups into the U.S. labor market created conditions that at times divided American workers in struggles over community resources. Katznelson's description of the formation of city trenches formed by values and customary practices, such as the deep-rooted idea that proper citizenship is intimately linked with individual home ownership, is an adequate way to understand workplace organizing and community electoral politics. Yet his separation of the space of political action into workplace and community does not account for either the importance of what I term liminal—intermediate— spaces or the discourse that is generated by urban spatial divisions that then construct the language of difference and belonging within American society. What is necessary is an analysis of a third space, the representational spaces by which urban residents utilize existing spatial structures and conceptions of space to redeploy the discourses of citizenship in an effort to achieve recognition. This approach recognizes that political action is not confined to the realm of workplace or community activism in an either/or dichotomy, nor does it assume that urban residents who labor for wages should, by some predetermined process, be more inclined to organize for large-scale political power based on economic class positions. Rather, it is an approach that recognizes that political action can take multiple paths, paths that are framed and contained by the built spatial environment. This was certainly the case for Middle West Side residents, whose retreat into defense of neighborhood enclaves frames their recreational, social, economic, and political activities. Although the retreat to urban trenches in Hell's Kitchen did not preclude unified political action, it often was the mediating force that prevented lasting solidarity.

Paraphrasing Marx, the labor geographer Andrew Herod states that "workers make their own geographies, but not under conditions of their own choosing." Recognizing the structuring power of the capitalist system of accumulation to create, through its need for constant spatial fixes, both the lived space and conceived space of the modern urban environment, Herod and other labor geographers have shown how workers utilize the opportunities existing within projects of urban spatial restructuring to promote their own political programs, though these are programs that do not necessarily comprise collective organizing or community activism. Herod states that urban workers counteract or redeploy the processes of urban restructuring when those processes are inadequate to their needs. "Herein lies a source of contradiction and potential struggle, for workers may have quite different geographical *visions* concerning how capitalism should be arranged spatially.

Consequently, workers may pursue a spatial fix either of their own organization or of capitalism more generally, which is quite different from that preferred" by those whose conceptions of the built environment tend to dominate its restructuring.[46] What Herod is pointing toward is the multiple ways residents of an area such as the Middle West Side form their own vision of how life under capitalism should be arranged through representational space. They thus act as citizens through the means of an already internalized spatial-temporal matrix.

This has numerous implications for both long- and short-term political sequences. In the short term, urban residents interpret spatial restructuring through their own cultural lens, and create or produce different kinds of space than that expected by conceptual spatial planners. This is the space I define as liminal, which is formed between the lived space of accumulation and the conceived space of restructuring, essentially as the door or entranceway from one to the other. As previously outlined, representational space refers to the space of perception, how space is understood. It is the interpretation of lived and conceived space, a process embedded in the cultural signs and symbols already produced by the environment. Though perceived or representational space will not automatically correspond to the needs and desires of those who have influenced, through their conceptions, the lived space of the urban environment (reformers, capitalists, government officials in the case of Progressive Era New York City), it will necessarily reflect those needs and desires. Yet representational space will also reflect the inability (whether planned or accidental)[47] of city planners, real estate investors, government officials, and other ideologically oriented groups to create or produce spaces that can fully capture and homogenize the spatiotemporal horizon, that is, the vision of urban collectives. To paraphrase Ernst Bloch, not all people live in the same space, even when they occupy similar locations.[48]

These implications lead to two questions of vital importance: first, how does representational space generate or constitute discourse, and is the generation individual or collective? Second, what are the dominant cultural frameworks in which representational space is embedded? Taking the latter first, for the period and area under study the physical lived space is dominated by the processes of uneven geographic development, and conceived space is framed by the needs of accumulation and reform. Thus the environment in which residents experience the everyday is one in which economic insecurity, the struggle for public resources, the construction of gender and gender roles, and the production of racial and ethnic difference deeply influence the way the built environment is perceived. These processes, struggles, and negotiations take place at the local level and within the wider context of city, state, and nation. As such, they are framed within the larger questions of legal status, economic rights, and political recognition. In other words, they are questions of belonging, of citizenship.

Struggles over resources, negotiations with civic authorities, decisions regarding the allocation of public and private investment, and the framing of racial, ethnic, and gender categories are all generally questions of belonging and citizenship. As political scientist Judith Shklar observed, "There is no notion more central to politics than citizenship, and none more variable in history or contested in theory." In the case of American citizenship, Shklar is one of many scholars who see the changing boundaries of U.S. citizenship as a continual struggle between exclusionary and inclusionary practices and methods. "The struggle for citizenship in America has, therefore, been overwhelmingly a demand for inclusion in the polity, an effort to break down excluding barriers to recognition," and thus a matter of both legal status and "public standing."[49] For Shklar, the twin emblems of American political belonging are the right to vote and the right to earn. Both rights are gained or withheld as matters of legal standing and as matters of discursive belonging. Her work in *American Citizenship* centers on the struggle of women and African Americans to achieve the legal status of voter and economic agent within an exclusionary legal framework that withheld these rights by statute and by tradition. For the purposes of studying the Middle West Side during the Progressive Era, when the legal status of residents was mixed in regard to the right to vote and earn,[50] the focus will be upon the discourse of citizenship, languages, and signs, which though connected to legal status in very real ways, are mainly concerned with matters of opinion, tradition, and perception. Even though the category of citizenship refers indirectly to the legal framework, it refers more directly to the perceptions and self-perceptions of both urban residents and those who sought to mold or direct their place within the political body. Put simply, citizenship here refers to the right, at the level of daily life, of individual and collective bodies to be viewed as full members of the larger community and to have their individual and collective needs respected and taken seriously by those in authority and by fellow citizens. In short, it is a question of belonging and recognition, one whose language is generated by the built spatial environment and produced within the larger context of national conceptions of political community.

The question of who qualifies as a full member of the political community is one of the most contested and contentious in American history. In two works, *Civic Ideals* and *Stories of Peoplehood*, political scientist Rogers Smith examines the history of American conceptions of citizenship and the shifting boundaries of political belonging.[51] In the former, Smith presents what he sees as the three competing traditions of American conceptions of citizenship: liberal, republican, and ascriptive. In the latter, he charts how political elites utilize these conceptions to organize constituencies based on shared traditions and assumptions. Smith constructs a model by which political actors attempt to shape the temporal horizons of a people by appealing to history (ethnic, racial, civic) and the present and projected future (economics) to generate a non-determinative causality that frames categories of belonging. These mythologies and stories, grounded in material conditions and perceptions, are, in Smith's analysis, imposed from above by those who possess the ability to control and manipulate the discourse of citizenship and belonging. Though he acknowledges that the construction of "political peoples" is a process of "constrained, asymmetrical" interaction between leaders and led, Smith leaves no doubt as to which group dominates the process. As he states, "It is actual and would-be leaders who most directly articulate and seek to institutionalize conceptions of political peoplehood."[52] Thus the work of ideology, I suggest, lies in this attempt to control the temporal horizons of the constituent body of the political people by constructing narratives of belonging and types of knowledge that explain time as linear, progressive, and unidirectional. Groups who had been previously excluded from the legitimate political body can enter into proper citizenship through acceptance of the common heritage of the past, assimilation of dominant values in the present, and proper future-oriented activities such as homeownership and educational training. Certainly the actions of Progressive Era reformers, viewed in this light, can be seen as multiple attempts to shape the temporal horizons of the discursively excluded (the urban poor, ethnic groups, immigrants) by providing them with the spatial environment that would create what social theorist and Progressive Simon Patten termed the "rational citizen." In an almost-too-perfect example of Smith's model, Progressives attempted to shape the rational citizen by constructing stories of proper action based on a temporal horizon that was linear, progressive, and unidirectional, grounded in the acceptance of a common heritage, assimilation of dominant values, and proper orientation toward the future. They sincerely hoped that the provision of a proper spatial environment, including park space, bathhouses, clean streets, rational interiors, and recreational space, would generate a citizen more in line with what they saw as the needs of a modernizing economy and complex society.

Any careful study of the Progressive Era and subsequent history reveals that the efforts of reformers created neither rational cities nor rational citizens. Conflicting conceptual visions among reformers and government officials, the workings of the private marketplace, and the needs and desires of residents, all contributed to the construction of a built environment that was unevenly developed and far from rational. To more fully understand what was created, it is necessary to examine the relationship between the built environment and representational space while, as theological philosopher Jonathan Boyarin urges, we "expand our peripheral vision a bit and consider some of our assumptions about time and space."[53] Boyarin suggests that our understanding of social processes are limited because of the assumption that these processes are played out within a Cartesian world of gridded, established knowledge of the spatiotemporal world, where time is linear, progressive, and unidirectional, and space is the geographic context in which the temporal unfolds. Boyarin promotes a different approach to the social understanding of the spatiotemporal matrix, one that incorporates contemporary theories of physics that view the human body/subject as not separated from the object world but ontologically embedded within it. This approach views space, understood here as the built environment, not as the product of human endeavor, but rather as an integral part of the larger process of the functioning physical environment. In other words, space is not just produced, but is producer as well, in a constant process that has nearly limitless implications for our understanding of the construction of social categories such as citizenship.

So what does space produce? I contend, as suggested earlier by Harvey, that the built spatial environment produces the very language and discourse that is used to evaluate its own construction. In other words, taking the Einsteinian view, space warps time to form the spatiotemporal matrix and the conditions of possibility for action. Thus as progressive reformers attempt to affect the temporal horizon of expectation of urban residents, those horizons are already grounded in the material existence of the built environment, and the languages of understanding, navigation, and negotiation produced within such space. The ideological vision of the urban planner (conceived space) and the lived space of market process and accumulation produce the modern temporality of crisis, in which the modern city is always in the process of restructuring, whereas the representational space of residents grounds this temporal horizon in the expectations of the everyday and the multidirectional experience of history, where past, present, and future become not the arrow of progress but the means of everyday imagining that translates into sometimes militant action.

Spatial restructuring produces alterations at three levels: lived, conceived, and representational space. These alterations in turn produce languages of understanding and knowledge that are contested and deployed by relevant social actors in the interstitial spaces that escape the plans of conceptual spatial experts, such as when residents reuse space intended for other reasons. The creation of liminal spaces does not necessarily produce discourses and understandings that are progressive, transgressive, or even necessarily political. But they do produce the altered representational spaces through which new and different understandings of categories such as citizenship emerge. It is within this process of the production of space that democracy, the ability to participate in the decision-making processes that affect daily life, are contained, contested, and interpreted while being embedded in the spatiotemporal matrix of restructuring and reform. New conceptions of citizenship, both at the level of outside observers and at the level of urban residents themselves, are contained within this matrix.

In the following chapter I trace the implications of this view of space and the restructuring of the built environment through an examination of a series of discrete events and their relation to events at other spatial scales. It is my hope that these implications will be useful and productive not only for a rethinking of the history of the U.S. Progressive Era, but for examining spatial change within a variety of scales and in a variety of historical settings. Guided by historically contingent forms of investment, production, distribution. and consumption, spatial change is a constant process, and is constantly producing different forms of lived, conceived. and representational space. Private investors' need for a spatial fix, combined with governmental action and use-value initiatives, work to perpetually restructure space, particularly at the urban scale. Urban residents' similar desire for a spatial fix works to destabilize processes of rationalization and market schemes to produce spaces that are contingent, unpredictable, and constitutive.

As Harvey's "difficult proposition" suggests, the way that resources are deployed in urban space is contingent upon the framing discourse produced within the space itself. Urban spatial restructuring produces knowledge about urban processes, including an epistemological approach to urban aggregate populations that serves not the needs, perceived or real, of that population, but rather operates within a dispersed web of power that encourages and inhibits certain practices. Not a new citizen—a rational citizen—emerges but a different collectivity, ceaselessly producing the spaces of urban life.

CHAPTER TWO

Restructuring Progressives

The domicile consisted of three extremely dirty and slovenly rooms in the upper part of a dingy tenement. It was presided over by a woman weighing perhaps 250 or 275 pounds, who smoked a pipe and smelled of beer, tobacco, and other things. She greeted the nurse in person, extending her hand, but not in a courteous salutation.

"Git t'ell out a' dia, before Oi smash yer bloody 'ed wit dis' rollin' pin," she bellowed, by way of polite rejoinder to the nurse's information that she had come to see Terrence.

"But I'm the nurse from the Health Department," began the visitor, Miss Adams, displaying her badge.

"G'wan now, g'wan before I hit ye hard," cried Mrs. McKay, "and may the devil go wit ye." [54]

THE PARTICIPANTS AND the setting of this exchange, drawn from a *New York Times* feature story of 1905, will be familiar not only to students and scholars of the Progressive Era, but also to museums dedicated to aspects of the "American experience," and, given the proliferation of historical documentaries on television, to the casual follower and hobbyist of U.S. history. As the historian David Hammack states, "The compelling images of Fifth Avenue and the tenement house reinforced the contemporary conception and have combined with the striking language of contemporary journalists and reformers to make New York of this period one of the most widely known, if not one of the most accurately understood, of

America's historic cities."[55] Miss Adams is in the familiar role of the reformer, the high-collared, twenty-something, middle-class, cosmopolitan social worker encountering the recalcitrant slum dweller. Mrs. McKay is a stock character as well, unkempt and suspicious, playing the role of the urban provincial,[56] locked into a local world of poverty and limited horizons. The two meet in a space where reality meets image, an apartment in a Middle West Side tenement whose interior and exterior are bound up in the lived space of the city and the conceptual space of the reformer, defined by a language of impending urban crisis.[57] "Dingy," "dark," "foul," "loud" evoke the sights, sounds, and smells of an almost trans-historical urban setting, framing the boundaries of social class, taste, and proper behavior.

The way we think about, write about, navigate through, conceive of, and plan for contemporary urban space is stubbornly infected by the language of encounter that derives from such scenarios as the meeting of Miss Adams and Mrs. McKay.[58] The coded language of danger, development, criminality, improvement, and proper citizenship is deeply embedded in the processes of uneven development that determines investment and restructuring in U.S. cities.[59] Neighborhoods are frequently described as dangerous, as if the bricks could leap from the walls to injure the unsuspecting passerby. Neighborhoods cannot be dangerous in and of themselves, as a cognitively or geographically bounded territory. What the coded language refers to is the residents of certain areas, who are perceived by outsiders as "us" or "other" based largely on the prevailing commonsense knowledge of the area of residence. Of course, spatially identified urban zones, divided by use, profession, and the relative societal wealth of residents, has a long history, maybe as long as the history of cities themselves.[60] Yet the particularities of the perception of urban space as a zone of impending crisis during the Progressive Era has a unique relationship to the development of modernity and a lasting effect on subsequent and contemporary views of the urban problem.

For the earnest progressive, from the anonymous Miss Adams, whose type is multiplied by the thousands, to the pantheon of U.S. Progressivism—Felix Adler, Jacob Riis, William Allen, and the like—the urban poor and the "slum" represented the textual space of common understanding where the latest theories of reform, social science, and moral environmentalism synthesized to form the

A tenement on 49th St. at Tenth Avenue with a brownstone front. (Milstein Division of United States History, The New York Public Library, Astor, Lenox and Tilden Foundations)

discourse of reform and a key period in the emergence of modern America.[61] It is no coincidence that in the period under study here the philanthropic Knickerbocker and the haughty Mugwump, figures of an older New York who believed urban problems were a matter of morality, are slowly superseded by the scientifically minded social workers, aided in their efforts by new levels of public visibility that go along with new publics.[62] Moving from ameliorative organizations such as the Association for the Improvement of the Conditions of the Poor to sophisticated and statistically obsessed Taylorist reform organizations such as the New York Bureau of Municipal Research (BMR) meant moving from viewing the poor as objects of inherent immorality to the "modern" belief in the connection between space and behavior. Coming to prominence after 1900, scientifically oriented reformers attempted, in the words of Henry Bruere of the BMR, to "utilize the same techniques as those employed by modern advertising" to sway a reluctant working class to abandon ethnic, religious, and political party ties for the pursuit of rationality in civic life.[63] The new levels of visibility were formed by the professionalization of the social sciences, the growth and sophistication (in terms of sales, price, and prurient appeal) of the tabloid press, the increasing reliance upon statistics, charts, and diagrams, and most important, the widespread use of the photographic image to convey increased power for written text.[64]

The synthesis of language, statistical compilation, scientific method, and the image formed the core of the Progressive worldview, bringing urban space into new focus and creating the indelible image of crisis. It was one thing for the upper- and middle-class urbanite to observe the surface image of the "dangerous" classes—poorly dressed, consumptive, scurrying to long days of labor and nights misspent in vice—but the world of these residents was geographically constructed to avoid intimate contact. It was quite another thing to be confronted with the physical evidence of the interior of urban poverty. The statistical compilation of disease and mortality, the charting and diagramming of overcrowding and poverty, and the reproduced images of "dingy" and fetid places created the basis for the discourse of temporal crisis derived from the materiality of urban space. Under conditions of mass democracy, the Progressive asked, what hope was there for a liberal, self-governing citizenry in the face of such ignorance and depravity? It was these uniquely modern modes of communication and understanding that spurred New York City Progressives to action, and helped to formulate their near-obsession on the linkage between space and physical and mental health.[65]

But this is not a history of the Progressive Era, or a re-asking of the question: Who were the Progressives? Nor is it a history of New York City. It is a historical account of the production of space at the urban scale in a particular city during a specific period, and an examination of the relationship between the production of space and common understandings of citizenship. But to understand this process, it is necessary to know not only the goals and motivations of Progressives, but also the epistemological systems they constructed regarding space and behavior. This understanding of what constitutes urban conceptual space allowed reformers such as Lawrence Vellier to speak confidently and without fear of rebuttal on questions of spatial restructuring: "The value of the small park in a tenement quarter is so well understood that it seems unnecessary at this time to present any arguments for the expansion of this work."[66] It is also necessary to compare the utter self-confidence and sense of common knowledge in such statements with the relative lack of real and lasting spatial improvements during the period. An exploration of the workings of city administration and dealings in the private marketplace can account for much of this lack of genuine accomplishment in the lived space of the city. And finally, the study must include the activities and actions of the relatively anonymous mass of urban residents—the urban poor, the objects of reform, the signifiers of crisis.

PROGRESSIVE VISION AND VISIBILITY

VISUAL IMAGES POSSESS an affective set of potentialities that differ from those of the spoken or written word. Vision, sight, is in its primary mode a spatial sense. One does not see time, but vision can create the sense of temporality in the spatial horizon. The photographs of Jacob Riis, which froze urban space in a represented moment, are said to have sparked the urban reform movement in New York City by showing the "true" conditions of urban poverty. They brought into the visible realm the darkest corners of social, physical, and cultural deprivation.[67] Though the enormous disparities of wealth and the physical conditions these disparities produced had been conveyed through written and oral testimony, and were widely but perhaps not deeply acknowledged, Riis's photographic essay, utilizing the latest in flash technology, created the sense of impending crisis that would dominate the Progressive Era. Yet in the immediate aftermath of this act of illumination, little was actually done to alleviate the worst conditions of New York's slums or of tenement dwellers.[68] The exposé of how the other half lives was an effective spur to some reforms, but the photos served other purposes as well. On the one hand, they formed a set of visual representations that made intelligible to the middle-class eye the seemingly irrational nature of the life and social activities of the urban poor. Along with voluminous written reports and statistical studies, and the learned writings of the leading figures of the emerging social sciences, the visual images formed the common understanding of the need for reform by making visible the spaces of vice, ill health, and poor habits. It was in photo studies like those of Riis and others that the visible image of the resigned laborer, passively accepting conditions unfit for normal or middle-class citizens, created and encouraged an emphasis on the next generation, the children, and the basis of urban crisis thinking. How could these younger ones, gaunt, careworn, aged beyond their years, be spared from the resigned fate of the older generation, and more important, in an age of expanding suffrage, how could they be turned from a dependent class to upstanding, self-governing citizens?

Secondly, these images—the photographs, charts, diagrams, and graphs, along with the physical remains of architectural structures—form the spatial archive of the built environment. Drawing on this archive allows us to scrutinize and unpack the Progressive Era epistemology of moral environmentalism and the near-obsession with spatiality that framed so much of their reforming zeal. The emphasis of urban reformers on space and health is well established in the historiography of the period, yet the relationship between Progressive discourse and the images they produced has barely been studied. What role did the actual physical spaces of the urban world and the images produced by new technologies and administrative techniques play in forming the common modes of communication shared by Progressives? How was the spatial imaginary, the very way reformers thought about urban space and conceptualized its role in shaping people's behaviors, formed by their encounters with the actual space itself? How was the reconceptualization of the Progressive imaginary embedded in the very images produced by these techniques? What did the image make visible?

Writing about reform in Victorian English cities, historian Christopher Otter suggests that the Progressive obsession with the space of the slum was bound up in the concept of visibility.[69] For the Victorian reformer, as for the U.S. Progressive, making the dark corners and narrow passages of the slum visible drew all that was uncivil into the open light of respectable distance. Proper self-governance could only take place when behavior was modified to be acceptable to the middle-class gaze, displayed for all to see and judge. Otter writes, "The respectable mastered their passions in public spaces conducive to the exercise of clear, controlled perception: wide open streets, squares and parks. In their homes, separate bedrooms and bathrooms precluded promiscuity and indecency. This can be termed the 'bourgeois visual environment.'"[70]

Restructuring Progressives 51

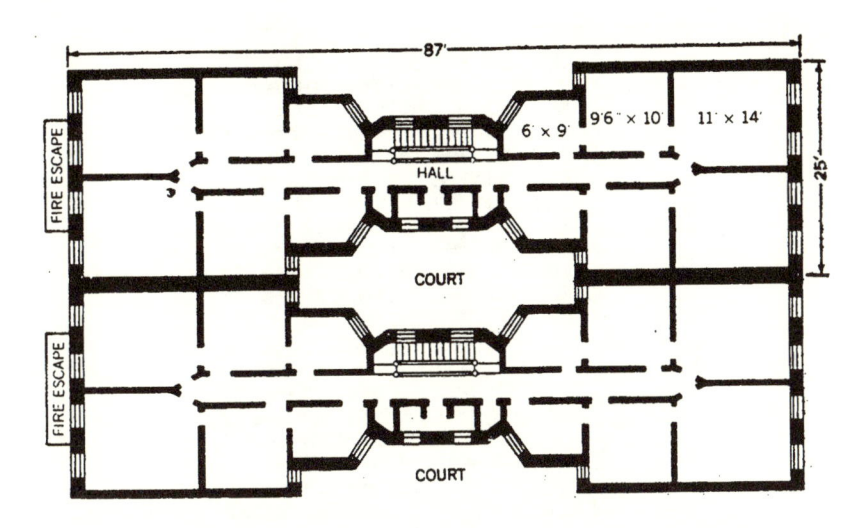

Design plan of the infamous dumbbell tenements that dominated Hell's Kitchen. (From *The Tenement Housing Problem*)

For the reformer, the space of the working-class neighborhood was the antithesis of this visual environment—crowded, cramped, with hidden corners, narrow alleys, overpopulated interiors, spaces that negated visuality, where conduct could not be seen and evaluated based on accepted codes of behavior and decorum. It stands to follow, as Otter states, that for the Victorian and the Progressive, the denizens of such environments "could not be trusted with social and political freedoms." So while cramped spaces without proper ventilation and dirty streets lacking maintenance bred disease, criminality, and mortality, they more importantly served as signifiers of a future crisis, particularly concerning the youth. How could these young people grow into proper citizenship, be trusted with liberal "freedoms," when their values were formed in such spaces?

Space, visibility, and health—physical, mental, and social—formed the holy trinity of the gospel of moral environmentalism and the basis for the Progressive obsession with reconceptualizing urban space. *Health* here refers not merely to the physical functioning of the body but to the proper maintenance and functioning of the community, at the local, city, and national level. Indeed, for the reformer, it referred to the future of democracy itself. The obsession with space and health framed Progressive ideology, particularly

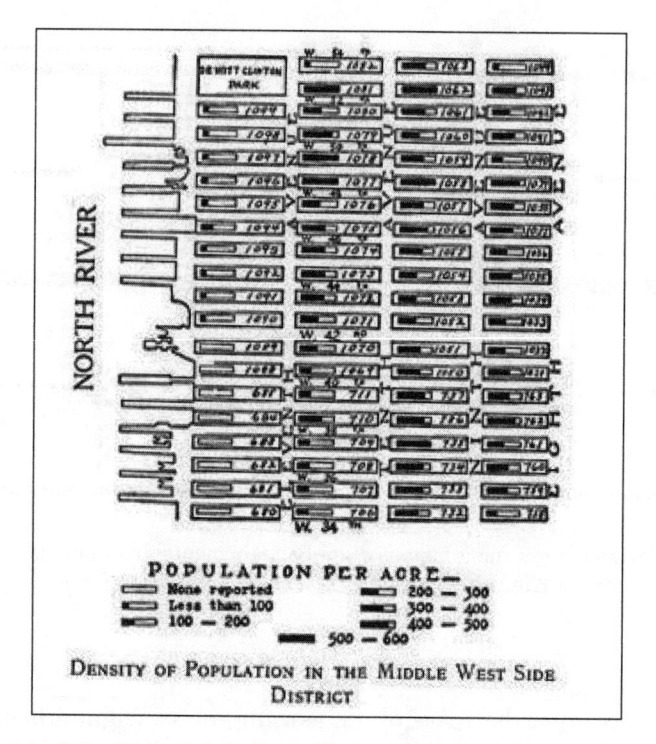

Figure 1A: Otho Cartwright's diagram of population density in the tenements of Hell's Kitchen, 1912. (From *The Middle West Side: A Historical Sketch*)

after 1900, and constructed a reconceptualized urban environment that could act as a palliative for the ills of mass democracy. Four images from the spatial archive illustrate the process of knowledge production and the sedimented construction of a common vision of conceptual space shared by urban Progressives.

The image pictured above and on the facing page are diagrams of urban space. Figure 1A is taken from Otho Cartwright's "The Middle West Side: A Historical Sketch," published in 1913. It is intended as an accurate depiction of the overcrowded conditions of Hell's Kitchen. Using information from census data and social research surveys, it is both a statistical and visual image of population and structural density. The numerical representation of the block-by-block population density of the neighborhood, revealing in some areas 500 to 600 residents per acre of land, would be useless and unintelligible to those without knowledge of urban

Restructuring Progressives

density problems or without the ability and common understanding to convert the pictorial representation into a mental image of overcrowding. Two things stand out regarding this image. First, Cartwright assumes his readership would understand not only the problem of physical density represented by the drawing and accompanying graph, but also the consequences, both spatial and temporal, of such conditions. Cartwright's diagram and graph form a language of image and number, quantification through tropes of visibility, which by 1913 had become a standard form of Progressive dialogue. Cartwright did not need to draw pictures of ill health, criminal acts, and improper behavior between his rectangular blocks. His audience, reformers and politicians on the reform bandwagon, implicitly understood this hieroglyphic representation of urban crisis.

The second striking aspect of the diagram is the space of the drawing itself. Meant to represent Eighth Avenue to the river between 34th and 54th Streets, the drawing is cramped. The letters demarcating the names of the north-south avenues span the entire width of the streets, as if large letters could span what was, in reality, very wide boulevards, albeit ones choked at times with traffic, pushcarts, and, on Eleventh Avenue, freight trains. The east-west

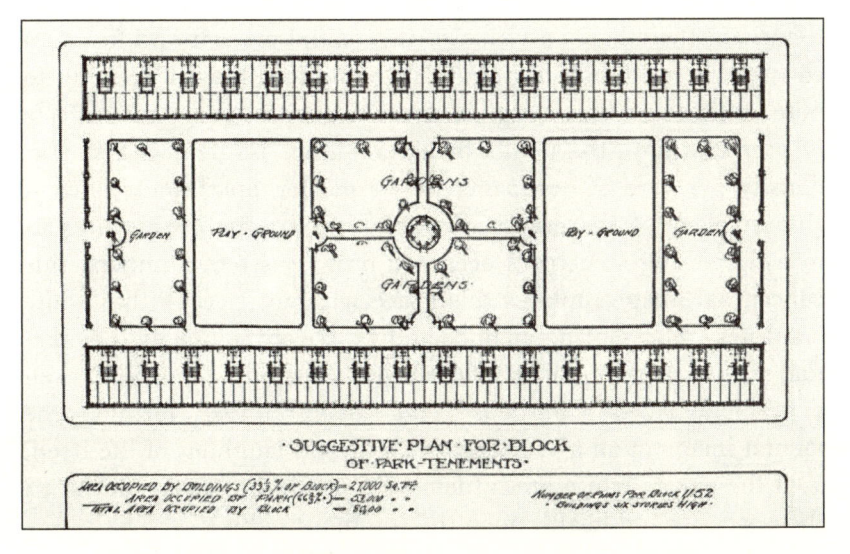

Figure 1B: A Suggestive Plan for Park Tenement Building.
(From *The Tenement Housing Problem*)

streets are equally narrow, with monolithic rectangles representing the built structures. Though each block contained different types of structures, often a mix of three- to five-story tenements, low-slung small factories, distributors, stores, and some two-story homes, the block drawing makes it appear claustrophobic, intelligible as overcrowding to even the untrained eye, a space without space.

Figure 1B, in comparison, is a "Suggestive Plan" for a block of garden apartments from Robert DeForest and Lawrence Vellier's *The Tenement House Problem*, published in 1903.[71] The image precedes a chapter in this volume, "A Plan for Tenements in Connection with a Municipal Park" by the architect and reformer I. N. Phelps Stokes.[72] Meant to depict an abstract "block" of urban space, this conceptual design sketches a vision of restructuring framed within the same set of assumptions that guide Cartwright's diagram, Figure 1A. The intimate connection between space, visibility, and health is obvious. The common uniformity of each diagram reveals the implicit assumptions of city planners and urban reformers. Following a shared master narrative, both show their reliance upon orderly lines, spatial demarcations, distance, and concepts of visibility and obscuration that are accompanied by numbers, charts, statistics, and labels. Though not assembled at the same time by the same sources, depictions such as Figures 1A and 1B communicate across distances by utilizing a communicative chain that shares a vision of the spatial present and future, a conceptual relationship of crisis and solution. It is interesting to note that though Cartwright is concerned with the negative effects of overcrowding, the model block of Figure 1B depicts a similar density per acre of occupants in the garden apartment concept. But in place of the cramped space between the geometric patterns depicting structures occupied only by street numbers, this conceptual image replaces this space without space with a white emptiness representing an open and visible environment. The textual structure of this blank space is occupied only by benign, even benevolent labels—"gardens" and "playgrounds," forming the mental image of an airy space, of children's laughter, of life itself, as if the empty white areas themselves could bestow health upon the residents and health upon the commonwealth.

This common language of spatiality, visibility, and health united men like Cartwright and Phelps Stokes. For both Cartwright, the middle-class reformer, and Phelps Stokes, the "Four Hundred" family architect and philanthropist,[73] visions of spatiality and conceptions of restructuring defined the terms of the present crisis and the solutions. The cramped drawing of Cartwright and the airy space of Phelps Stokes's model block contain both the painstaking lines, graphs, and accumulated knowledge of the empirical science of reform and the unseen, unrepresented—yet implicitly understood— objects of reform. What is implicit is added by the photographic image. Figures 1C and 1D (pages 56 and 57) typify the imagery of crisis of reform language. In Pauline Goldmark's *Boyhood and Lawlessness*,[74] as with many reformist works, amelioration for the present generation of adults was only the most basic, and perhaps least important, aspect of the reform agenda. Restructuring of urban space is the barest palliative for the present generation. What Goldmark's photographic images add to the statistics, graphs, and diagrams is the threatening image of a future of mass democracy without self-governing citizens. Figures 1C and 1D add the element that is unspoken in the representative language of Cartwright and Phelps Stokes but surely implicit in the reformist understanding.

The master narrative of urban reform, the standard plot of the characters depicted in the opening of this chapter, revolves around the relationship between past, present, and future embedded in spatiality and visibility. Miss Adams, our reformer, must penetrate the labyrinth of the tenement and make it past the helpless Mrs. McKay to find the child and bring him out into the light of health and education. The shock value of Riis's photo essays, the sense of impending crisis in the voluminous reports of reformers, the didactic quality of the graphs and charts—all rely upon an implicit understanding regarding the threat posed by urban poverty. *Boyhood and Lawlessness* shares a similar plot structure and communicative chain with other narratives of reform: economic hardship and a lack of personal qualities combine to form a "degraded" humanity in the urban slum, where adults are dragged down by life's circumstances, accept their lot, become a "people without wants," and pass on this "bitter lesson of endurance" to their children.[75] In the narrative, urban space becomes the vehicle for this temporal sequence. The threat here is hardly implicit, but starkly explicit, represented in the photographic images of crisis and reform. The past, the circumstances of poverty, creates the spatial present of the dark corners,

Figure 1C: "A 'Den' Under the Dock."
(From *West Side Studies: Boyhood and Lawlessness*)

narrow alleys, hidden recesses that allow the indecent, the diseased, the improper, to thrive, creating a present of no future for the urban poor, and a threatening future for the Republic. To break the temporal chain, to interrupt the flow of acceptance and endurance, the space must be restructured to expose the improper to the naked light of public judgment.

Figures 1C and 1D speak both to the image of future crisis and the conceptual image of restructuring. The photographic images work in combination with the text and the established discourse of space and health to connect past, present, and future into a temporal continuity that requires the restructuring of the spatial. For the reformer, the past produces the haphazard, random environment of the un-zoned neighborhood, where the demands of industry and economy, though seldom directly held to blame, create the conditions of non-visibility that allow for the maintenance of dark, hidden spaces of vice, immorality, and disease. Decaying waterfront docks, basements, unlit hallways are the conceptual present that prevents the proper nourishment of body and mind, guiding the young as a spatial Pied Piper to the underground world of the unofficial economy, the space of easy money, low morals, and licentiousness. The built space enables the easy transition from childhood innocence to gang membership, a fundamental inevitability of Progressive discourse played out on the physical grounds of the overcrowded,

Restructuring Progressives 57

dank-smelling slum. All of the images project into the future, anticipating the crisis that can only be halted through spatial discontinuity, a reconceptualization of urban space.

In figure 1C, with its accompanying text, "A 'Den' Under the Dock," one can almost feel the fear in the use of quotations around the word *den*. The play of light and shadow reveals the progressive emphasis on visibility. The faces of the boys suggest a playful mischief, but the source of their contentment remains unseen. For these boys, Goldmark writes, occluded spaces are the "playgrounds" of Hell's Kitchen, the place where there is always "somethin' goin' on," the locale of "selfish and reckless" behavior. Goldmark's *Boyhood and Lawlessness* contains many other photos of such neighborhood spaces, where boys shoot dice, hold impromptu and unsupervised boxing matches, and divvy up their ill-gotten loot. Yet this picture is particularly telling as a visual image of impending crisis when considered along with one of the hallmark tropes of reformist writing: the unsaid consequence of unsupervised, invisible behavior. This manner of writing, found in numerous texts supporting reform, acquires evidence regarding the causes of improper behavior, building to a crescendo of impending crisis beyond the everyday behaviors

Figure 1D: One Diversion of the Older Boys.
(From *West Side Studies: Boyhood and Lawlessness*)

observed in the streets. For Pauline Goldmark, unsupervised play in "sewage-laden" waterfronts and empty basements lures young boys into petty criminality, "or worse."[76] For Dr. Felix Adler, the lure of having one's own money in areas of prostitution steers young girls with "the strongest temptations" toward unsaid acts of immorality. Indeed, in the scenario that opens this chapter, Mrs. McKay smells of "beer, tobacco, and other things," the last being unspeakable and left to the prurient imagination.

The photographic image of Figure 1C forms an interesting and perhaps telling parallel with the sense of impending crisis related to both the unseen and unsaid. The source of the boys' contentment lay in their access to the dark space beneath the dock, out of sight of parents and police. The photograph captures the Progressives' sense of fear surrounding the lack of visibility. What cannot be seen cannot be exposed to the light of reason and proper behavior without spatial restructuring. If boys are allowed to inhabit such spaces, and play unsupervised, they may become attracted to those things that lurk in the unseen corners, whether it be petty criminality, sloth, slovenliness, "or worse." The "or worse" is captured in Goldmark's lurid and perhaps fantastic descriptions of tenement sexuality, which include descriptions of infant sex play and toddlers engaged in sodomy.[77] This connection between invisibility, urban space, and improper behavior is further illustrated in Figure 1D, with its corresponding caption of erotic panic, "One Diversion of the Older Boys." The sexual connotations convey in the image of the endangered child the future-oriented nature of urban crisis thinking.

Figure 1D captures a typical street scene of the MWS (possibly Tenth or Eleventh Avenue, the photo credit does not specify), and more stock images of reformist discourse. The boy, who appears on close examination to be about fourteen or fifteen, is a classic "gopher," a budding gang member on the verge of true criminality. The two girls, dressed in rather plain clothes, without hats or collars, are the very image of "tough girls," those without the factory or service jobs that provide independence and spend their time on the streets seeking amusement in a variety of forms. From the natural light of the picture, it appears to be either morning or midday, a time when respectable people are working or children are in school. The boy, with hands in pockets and slumping posture, seeks his diversion with the two tough girls, all three driven to the streets by the cramped conditions of the tenement interior, in enforced idleness because of either a lack of economic opportunity or from failed supervision. They seek their diversions within an environment that has failed to offer them alternatives and failed to properly socialize them into appropriate gender and sexual roles.

Though they are the center of attention, our trio of youths are not the only things visible in the street scene. Other negative aspects of the tenement district surround them. To their right, a grown man appears to be lying on a bin, looking disheveled, perhaps intoxicated. For Goldmark, the example set by careworn and beaten-down adults in the tenements sets the negative example for boys seeking diversion, such as this one. The crowded conditions of the domestic interiors, the prevalence of miscreants on the streets, and the general violent and immoral nature of many residents serve, as Goldmark states, to "strangle the sense of modesty" at an early age.[78] Pictured as well is a pushcart vendor, one of the many who contribute to the crowding of the avenues. Behind him, the street appears to be either undergoing repair (a rarity) or simply strewn with refuse, a more common occurrence. The avenue is broad, yet there is little sense of shelter; no trees are apparent, the appearance of the buildings is drab and monotonous, and one can almost feel the unrelenting heat of a hot summer day, almost smell the multiple meals being cooked, the refuse rotting. For many middle-class reformers, such streets constituted the antithesis of proper dwellings and neighborhoods.

Reformers like Goldmark equated the spatial environment pictured here with an almost cartoonish image of sexual license, which they linked to the impending urban crisis of generational decay. Goldmark paints a depressing picture of the diversions of the slum-dwelling boy, whose morality has been formed, or rather de-formed, within the confines of the invisible and occluded spaces of the tenement district. Here, on the rooftops, in the alleys and doorways, young boys learn that sexual enjoyment comes without responsibility. In the tenements, Goldmark writes, "immoral practices are common even among young children. All round him he is accustomed to hearing obscene terms spoken freely and robbed of any moral significance." The boys take these lessons, learned in the cramped proximity of the improper tenement interior, into the streets, where they pick up "the knowledge of perverts" that leads, because of the easy access to occluded spaces, to "experimentation" and "many forms of perversion" such as "sodomy." Ruth True, in her book *The Neglected Girl*, is even more spatially explicit. "The tough girls have two universal amusement places—the streets and the 'nickel dumps' (moving picture shows). Besides, they can make meeting places of alleys, the docks and vacant rooms in tenements. These neglected, unlit cracks and crannies serve as traps for both sexes. Here children are snagged in the darkness long before they are old enough to know the meaning of temptation. This is the most sinister phase of the recreation problem."[79]

Goldmark, True, Vellier, Cartwright, Phelps Stokes, and other Progressives constructed a common language of urban crisis that relied upon shared understandings of the images, statistics, and texts that portrayed for them the true nature of the crisis facing the Republic. They further shared common narrative forms that are familiar to readers of reformist reports: the objective description, the statement of the problem and the mounting of evidence, through both numbers and anecdote, usually followed by some suggestions on how to improve conditions. It is interesting to note that both Goldmark and True, in their treatment of tenement boys and girls, follow the same narrative framework, one that concludes with a discussion of sexual immorality. The echoes of this narrative of urban crisis are apparent in contemporary writing on the "problems of inner cities."[80] Improper environments lead to a host of ills that culminate in out-of-wedlock births that perpetuate the continuation of lineages of improper citizenship and threaten the self-governing nature of the Republic. Only by bringing the light of rational judgment to these occluded spaces through restructuring can the city be spared.

What was to be done to avert the urban crisis, to save the system of self-governing democratic citizens from the growing masses of the working class and the influx of immigrants? In his seminal work on the Progressive Era, *The Search for Order*, Robert Weibe sets the template for further research on the period in his explication of the shift from traditional "autonomous communities" to the "techniques" of a cosmopolitan middle class attempting to deal with the disjuncture caused by a new industrial and social order.[81] Weibe's influence can be seen in current surveys of the period, such as Robert Peterson's *A Fierce Discontent*, which view the Progressive Era as a largely class-defined phenomenon, an attempt to form a "paradise of middle-class rationality."[82] M. Christine Boyer takes Weibe's disjuncture in industrial and social order to a Foucauldian level, seeing in the rise of middle-class rationality the development of modes of disciplinary control that are internalized by citizens through daily contact with a new "infrastructural framework and regulatory land order" meant to construct the "rational city." Boyer's work makes explicit a spatially determined basis of reform: "Environmental reform was the most important disciplinary order upon which the new civilization of cities would rise."[83]

Any quick survey of the efforts and philosophy of Progressive reformers, and particularly those who worked in New York City, illustrates the spatial nature of their efforts and accomplishments. Between 1880 and 1920, the city saw the establishment of a small parks commission, the construction of playgrounds and bathhouses, new tenement laws, and various efforts to open up the cramped spaces of urban working-class neighborhoods to the visibility of middle-class moral judgment. Yet in spite of the intensity of their efforts and their well-intentioned attempts, the results for the city were in general mixed, and for the Middle West Side must be considered a failure. Despite the concerted efforts of reformers to bring the light of rationality to Hell's Kitchen, the neighborhood, though changed, would persist in the minds of New Yorkers as a "dangerous" area until recent decades. This is not to say that the efforts of reformers had no effect on either the space of the Middle West Side or the actions of its residents. But the effort to bring visibility to the area is also the story of the forces of occlusion, private economic interests, and political figures that worked often in opposition but sometimes in concert with reformers. It was the synthesis of these often competing forces that formed the "daily round" of the Kitchen's residents—the lived space of the everyday. It is the story of occlusion and visibility, of uneven development in the production of space.

VISIBILITY AND OCCLUSION

SPACE IN MANHATTAN above 34th Street was defined in the second half of the nineteenth century by the construction of two massive public parks, Central Park from 59th Street north, and Riverside

Park, hugging the Hudson River from 72nd Street north to 112th.[84] The area between these two famous greenswards would hence be known as the "Upper West Side," with its connotations of wealth, gentility, and sophistication. The construction of these public spaces, in the 1850s and 1870s, respectively, substantially increased the value of surrounding real estate and served to entrench economic interests dedicated to maintaining the rising value of their properties. The maintenance of these entrenched interests goes a long way toward explaining how the "daily round" of lived space encountered by the Middle West Side resident was formed. The story of uneven development and restructuring during the Progressive Era in New York is as much a story of economic interests, market forces, and political power as it is a story of urban reform.

A story about trash, involving one of the heroic figures of New York's Progressive Era, graphically illustrates the problems of uneven development, political and social power, and visibility and occlusion. An 1895 magazine profile extols the virtues of the area of the West Side comprising Riverside Drive and Park, making it seem as if the Creator himself had endowed this area as an affluent enclave of increasing property values. The park and drive "are so magnificently situated on the Hudson River along the bluff that skirts it that it could not be otherwise. Nature evidently meant the locality to be a great park, and nature, as usual, builded very well."[85] By 1895, Riverside Park was a well-tended and jealously guarded area of wealth. The same profile states: "A rise in the value of property on the west side has naturally accompanied its development, and this increase has by no means ceased." Of course, the construction of Riverside Park and the restrictive conditions that governed building construction in the adjacent areas was neither an act of nature nor of divine provenance, but rather the intentional actions of those whose investment interests in the area led them to advocate for city services, public financing, and restrictive regulations that well protected their entrenched interests. In 1895, the leading advocate for the maintenance of property values in the area was the West End Association, the successor group to the old West Side Association that had been established in 1870 to "promote public improvements north of 59th Street."

The West End Association was led, in 1895, by Cyrus Clark, a well-known figure in New York City financial circles and president of the Hamilton Bank. Clark and the organization actively worked to enforce formal and informal regulations barring commercial businesses from the immediate area, often encouraging wealthy residents to "buy out" prospective commercial interests in what amounted to a legal form of real estate blackmail. They also lobbied the city to allocate funds for improvements at a time when many of the reformers were fighting for city money to be used for park and bathhouse construction. The magazine profile indicates Mr. Clark was successful in his endeavors: "He has done efficient service in bringing about the embellishment of the district and in putting a stop to features that were objectionable." While the Central Park Conservancy and other groups protected the entrenched interests of property values around the west side of that space, Cyrus Clark and the West End Association did the same for Riverside Park's wealthy residents. In a telling episode, Clark extols his latest triumph, convincing the Board of Apportionment, the New York City governing body that controlled revenue allocation, to "appropriate a sufficient fund" to cart away the refuse lying in a dump at the foot of West Seventy-ninth Street. Essential to this appropriation was the presence of Colonel George Waring as the new head of the Department of Sanitation. Waring, a Civil War hero and dedicated civil servant, had recently reorganized the sanitation workers of the city, dressing them out in white uniforms, instilling military discipline, and winning the confidence of those who controlled the city's tax money. Waring's men efficiently removed the refuse heap from the Riverside Park area, efficiently utilizing the city's allocation. The refuse was deposited in an existing dump area at the foot of West Forty-second Street, in the heart of Hell's Kitchen.[86]

The depositing of the Upper West Side trash in the Middle West Side clearly demonstrates the uneven relationships of power at the level of city politics at the turn of the century and the problems associated with the reformist conception of urban space. While urban reformers were concentrating their considerable efforts to produce visibility in the underserved areas of the city, powerful forces were at work to occlude the vast pockets of poverty existing in the city. This process of occlusion took place at two levels. On the one hand, uneven development and the efforts to maintain rising property values in certain areas ensured that neighborhoods and regions like the Middle West Side were de-linked from the major networks of development in the city and persisted as pockets of poverty. On the other hand, the system of negotiation, debate, and struggle that developed in order to address questions of fiscal allocation and provision of services was dominated by institutions such as the West End Association, the Citizens League, the Union Club, the Association for the Improvement of the Condition of the Poor, and others. These institutions utilized their considerable political and financial power to draw together assemblages and rules, such as expert commissions, zoning laws, and reformist forms of political participation, to protect their entrenched interests while also seeming to serve the greater public.

In his study of power in New York at the turn of the nineteenth century, David Hammack defined power based on Max Weber's formulation, where certain individual or group actors "realize their own will" against various forms of resistance. Hammack's study, though valuable, was limited to the physical manifestations of power exhibited in the selection of mayoral candidates, the planning for subways and other infrastructure works, and political issues such as the consolidation plan of 1898 and the centralization of control over New York's public schools. His work focuses on the relative political influence of a variety of groups representing differing views on the "urban problem." What Hammack and others have studied are the varying relationships between fiscal, cultural, and identity interests and the ideological formulations used to back each such interest in struggles over urban restructuring. For Hammack, the struggle between Tammany loyalists and "Swallowtails," labor leaders and laissez-faire Republicans, reformers and supporters of local control, was a battle for supremacy played out between relatively equal political and economic interests, where those groups who can best mobilize both elite and mass support prevail.[87]

As Hammack has shown, political and decision-making power, and thus the power to shape lived space, was not controlled completely by any single group of elites, nor was it dominated by party "machines." His study of the competing interests that set the tone and terms of late nineteenth-century negotiations describes how wealthy business interests, as an example, backed both the "corrupt" ward bosses of Tammany and the "upright" reformers who were their electoral opposition. Protecting their various and often intertwined interests, "the very rich did not form a unified and effective power elite," and were further limited by "the considerable and increasing resources available to the less affluent." Thus decisions regarding the allocation of resources and the provision of city services, though not a simple matter of machine versus reform or capital versus labor, were publicly and privately debated, negotiated, and taken within the context of an institutional framework that often preempted the interests of the less powerful by setting the very terms in which the discussion would be carried out. Often these terms were constituted by the spatial arrangements of uneven geographic development that determined the lived space of the city, as residents of more affluent sections exerted more influence over decisions. A brief review of two allocative and regulatory processes, the construction of small parks and the reform of tenement laws, illustrates the processes of occlusion that worked to maintain uneven geographic development.

According to the principles of moral environmentalism, providing proper space to the city's poor population was an absolute priority. This provisioning, according to reformers, required the construction of proper interiors and exteriors. The Small Parks Commission and the Tenement House Committee, established respectively in 1897 and 1898, were formed to specifically address these issues. The need for small parks in crowded neighborhoods and the need to enforce standards in the design of tenement houses were clearly acknowledged by most of the city's political and economic leaders.[88] Yet in both cases, preexisting conditions, entrenched interests, and the limited spatiality of Manhattan had hindered most efforts to provide these needed improvements. In the case of park construction, the division of the city above 14th Street into 25- by 100-foot lots, the movement of population "uptown" spurred by rail and trolley travel, and the construction of Central and Riverside Parks left little space available for new projects. Installing even small parks in the city's overcrowded neighborhoods required destruction of existing buildings, compensation to previous owners, allocation of city tax funding to pay for the purchase of land, clearance, grading, and construction, or turning over such projects to private charities, which were faced with the same problems but lacked even the city's limited power of eminent domain. Thus any new project required the participation of landowners, ward politicians, the mayor's office, business interests in the adjacent areas, private organizations and advocacy groups, state politicians, and competing political party interests.

The same problems encumbered efforts at tenement reform, adding in politicians and businessmen vehemently opposed to government interference in what they viewed as purely market transactions, and the equally fierce opposition of landlords and investors, who insisted on minimum profit returns on their property holdings. As a result, nearly all decisions on such issues as tenement reform and park construction needed to be painstakingly pushed through a labyrinth of competing power structures.

By the mid-1890s, the city was undergoing yet another period of political transition, and the mantle of reform was claimed by all but the most aggressive and obstinate ward bosses. Following the Tweed Ring scandal of the 1870s, political power in the city had been contested between many groups. The remnant of old-line Mugwumps, mainly connected to the state Republican Party, battled for supremacy with the so-called Swallowtail Democrats—so known because of their preference for formal attire and aspirations to upward social mobility. By the 1890s, Tammany Hall, disgraced by Tweed, had reasserted its power by forging alliances with certain wealthy businessmen presenting themselves as a legitimate force in New York politics. Tammany competed with the Apollo Hall club, the Irving Hall organization (and in 1898, the newly formed McManus Club) for influence with the city's growing Democratic constituency.[89] In the wake of muckraking work like that of Riis, and the growing problems of overcrowding, all competing political factions, as well as influential private organizations, had, in some form, begun to address the "urban problem." For most, by 1897, this meant adopting the platform of reform in its various modes, from ballot reform to civil service rules, to the establishment of settlement houses and the advocacy for poor relief.

Establishing reform credentials meant taking on certain performative modes and adopting a particular set of ideas, whether they translated into action or not. Reformers all publicly claimed to be nonpartisan and to be familiar with the language of science or expert opinion. One way to accomplish this performance was to promote or take part in the establishment of a "commission" or "committee," where issues of the day could be properly studied over a period of time through reliance on expert testimony, statistics, and careful, considered judgment.[90] Thus in response to the necessities of a growing city, Mayor William L. Strong formed the Small Parks

Commission in 1897 as an advisory body to study the best way to provide the city's overcrowded regions with open park space. The following year, the Charity Organization Society established the Tenement House Committee, whose mission was to study the needs of the urban poor and make recommendations to the city administration regarding what new ordinances were necessary to secure decent housing for this population.

By the time of the establishment of the Small Parks Commission and the Tenement House Committee, the movement for building park space and for reforming tenement laws had a long history, albeit one of little actual accomplishment. In the case of parks, private charity organizations had demanded open space in the "Mulberry Bend" district of lower Manhattan as early as 1845. Organized advocacy for small parks formed around the creation of Central Park in the 1850s, as advocates for the poor objected to the new park's location, inaccessibility, and consumption of city funds. But it was the Tenement House Commission of 1884, appointed by Mayor Franklin Edson, candidate of both the Swallow-tails and Tammany, that was instrumental in passage of the 1887 Small Parks Act. The act authorized the Board of Street Opening and Improvement "to select, locate, and lay out such and so many parks in the city south of 155th Street as the said board may from time to time determine."[91] The act also allocated "no more than $1,000,000 be spent in any one year," an allocation that was typical of the style of city budgeting in the period, when many allocations were ad hoc, and accountability for spending was well below the common standards of modern budgetary procedures that would be adopted by New York and other cities before 1920.

By 1895, eight years after the Small Parks Act's passage, the Mulberry Bend park project, first suggested some fifty years earlier, had been completed. Disputes over land seizures, contracting for construction, and a general mistrust of city administration by powerful private groups led to endless delays in the projected park plans. It was not until 1895, with the passage of specific bills by the state legislature, that construction on two new parks in Lower Manhattan commenced. In 1901, as reported by Veiller and DeForest, three parks had been proposed for the West Side above 26th Street, but no actual work had commenced.[92] It would not be until 1902 that construction would begin on DeWitt Clinton Park in the Middle West Side, again the result of specific legislation.[93] The delays and lack of construction were often the result of the variety of influences mentioned previously. Landowners whose properties were seized would seek redress in court for fair compensation. Construction and purchase of land would be held up by politicians of opposing parties, often on the claim that contracts would be awarded to construction firms at inflated rates that included kickbacks, a charge particularly leveled at Tammany administrations. Inaction can also be traced to the instability of year-to-year budgets, which were based on previous years' allocations, making it difficult to plan for future needs. The formation of the Small Parks Committee in 1897, amid a seemingly endless array of committees and commissions, was an attempt to overcome these barriers through a reliance on the exhaustive report and alleged impartiality of participants.

The struggle over tenements was even more complex, with a new committee or commission seemingly appointed by every incoming mayoral administration after 1850. Although "slum" districts such as the notorious Five Points were identified before 1850, overcrowding in Manhattan became a major issue in the second half of the century, as the geography of the island combined with increased immigration, the grid plan of 1811, and the search for profit to create the congested conditions dissected by urban reformers. Congested and unsafe conditions had led to the creation of a buildings department, overseen by an appointed commissioner and operating within the New York Fire Department. The early Buildings Department and its superintendent had little independent power and produced nothing in the way of enforceable standards in the construction and maintenance of tenements, and the job often went to a political appointee as part of the extensive patronage system. The first law governing tenement construction, the Tenement House Act of 1867, was, like subsequent acts in 1879, 1887, 1895, and 1901, largely the result of advocacy by independent organizations combined with reports of government committees. In the case of the 1867 legislation, a committee formed by the State Assembly adopted many of the recommendations of the Council of Hygiene and Public Health, which had been formed by the privately run Citizens Association of New York. But as Richard Plunz points out in his study of housing legislation in New York City, "The mere existence of a building bureaucracy did not guarantee enforcement of the law."[94] Tenement laws and building legislation, meant to provide the openings and spaces of visibility demanded by reformers, often failed because of lax enforcement, political dealmaking and graft, and the relative influence of reformers on the existing city administration. Uneven enforcement served to hide rather than make visible the severe housing problems affecting the city.

Private developers were yet another source of occlusion in the effort to bring visibility to overcrowded neighborhoods. Each new regulation or suggested law concerning tenements resulted in a squeeze on the profits of private developers, who were largely restricted to building on existing 25 by 100 foot lots. As the city passed legislation requiring ventilation, light, running water, and private baths, landlords—oftentimes sublessees who turned a small profit margin—responded with attempts to meet the minimum requirements, or evade them, while seeking compensation in courts for any costs incurred. Private owners believed that they provided a service in the open market, and that conditions should be determined by private contract, and not the "interference" of outside experts, whose opinions they often disparaged. Responding to criticism of private landlords by the architect Ernest Flagg, and to proposals for model tenements that were impossible within the grid design of the 25- by 100-foot lot, tenement owner Peter Herter responds with barely concealed contempt:

> Mr. Flagg says in effect that all these men [referring to landlords] for all these years have been working, so to speak, willfully, or, as it were, in the air, making plans just to see how unsanitary and uncomfortable tenement houses can be. No. Our architects and builders have not been dealing with an imaginary problem, but one that contains a great many hard factors. They have had to deal with financial conditions, with the price of lots, the value of money, the taste of tenants, the desires and requirements of tenants, the several laws, the different municipal departments, and so forth. The pressure of all these factors has gone to make our tenement houses just what they are today. These are the forces that have created the tenement and controlled its evolution.[95]

In the perpetual struggle between landlords and reformers, owners of tenements seeking redress or relief from building code legislation often found a sympathetic ear in a legal system that leaned toward the protection of private property and market mechanisms. Legal challenges ranged from minor disputes over compensation to owners for meeting required improvements to complicated suits that challenged the very basis of the city's right to regulate housing. As Richard Plunz states: "The lack of substantial new legislation and the recalcitrant atmosphere for enforcement was partially caused by the threat of legal recourse on the part of anti-reform interests."[96] And these fears were indeed warranted. Striking down the major provisions of the 1884 Tenement House Law, the New York State Court of Appeals is unambiguous in its defense of property and the private marketplace: "Such governmental interferences disturb the normal adjustments of the social fabric, and usually derange the delicate and complicated machinery of industry and cause a score of evils while attempting the removal of one."[97]

The complicated and disputed process of tenement reform was affected by other factors as well, including state and national politics. Because New York was a key electoral state, national politics played a large role in determining candidates for the governor's office and determining party political platforms at the state level. The State Assembly and Senate, often controlled in the post–Civil War period by upstate Republicans, whose "machine" was every bit as effective as any city organization, had long battled with New York City officials for control over issues such as the police department, taxation, allocation of funding, and infrastructure construction. State Republicans, often with reason, considered most city administrations to be rife with corruption, and were thus hesitant to approve large-scale projects and their concomitant budget allocations.[98] Edward Marshall had summed up the question of city involvement in housing in *Century Magazine* in 1895: "During the existence of the present unstable and often corrupt systems of American municipal government it would probably be unwise to advocate the city construction or management of dwellings for the poor." Upstate Republicans were also hesitant to support government-sponsored housing plans for fear of "city administrations stacking municipal tenements with voters, ensuring political support."[99]

The disparate interest groups that contested the issues of park construction and tenement reform made the work of committees and commissions, with their airs of neutrality and nonpartisanship, a necessity in urban restructuring. Committees and commissions came in two forms: those formed by non-governmental advocacy associations, like the Charity Organization Society, and those appointed by state and city government. In the usual process, private groups funded by concerned, mainly wealthy citizens used the weapon of the exhaustive report or survey combined with publicity of the problem at hand, to pressure the state or city into launching its own investigation for the purposes of proposing legislation. While many, though not all, of the privately formed groups consisted of committed advocates for reform, city and state committees, in order to demonstrate neutrality, consisted of a mixed membership representing reform and the private market. Thus in 1898, the city-appointed committee on new building codes consisted of five Tammany Buildings Department appointees, three owner-developers, an iron contractor, an architect, and a corporation counsel, ensuring that entrenched interests would have a loud voice. Other committees appointed by both city and state government often had overlapping memberships with private groups and other committees, but attempted to avoid charges of bias by appointing equal numbers from all interested camps. Most of the members of Mayor Strong's Small Parks Committee of 1897, including ex-mayor Abram Hewitt and muckraker Jacob Riis, had served either city government or private committees, and each was appointed to appease a different interest group. As well, the formation of committees and commissions were influenced by private associations formed to protect the vested interests of small groups of investors. Organizations such as the West End Association exerted considerable political influence and often placed members on committees charged with studying matters of "public interest."

The result of the power struggles among the various interest groups concerned with parks and housing meant that, in large part, efforts to restructure urban neighborhoods were contained within the shared language of committees and commissions. Much as reformers and their organizations developed a shared set of assumptions based upon shared images and texts concerning the urban poor, government and independent committees also developed a shared understanding of restructuring based upon an institutionalized discourse. Although often incorporating much of the rhetoric of reform, these institutions, both governmental and private, organized the committees and rules for debating questions of urban restructuring. Governmental and private committees and commissions incorporated the ideas of architects and engineers in their studies through design competitions and consultations. They brought urban reformers into their processes by adopting the rhetoric if not the reality of the reform program, often through an acknowledgment of the problem and avoidance of the suggested solutions. As they became nearly the only acceptable institutional form for organizing the variety of interest groups promoting or opposing restructuring, committees and commissions integrated architects, engineers, scientific reformers, moralists, nativists, multicultural cosmopolitans, machine politicians, career bureaucrats, and others into their linguistic universe, one whose rules were based on existing spatial conditions and established rights of property.

The rules governing the establishment and work of committees and commissions reflected the changes in both the physical composition of the city and the ideas formed about these urban conditions. In the thirty years following the Civil War, private and government committees were dominated by the language of morality as related to individual behavior. Organizations such as the Association for the Improvement of the Condition of the Poor and government committees like the 1884 Tenement House Committee shared not only common membership, but also a shared view of the subjects of their studies. Charles F. Wingate, a member of both institutions, expressed a common understanding of the urban problem in his description of immigrant groups as "ignorant, filthy, and more or less debased." For "reformers" like Wingate, morality was a matter of national origin combined with conditions of decrepitude.[100] Wingate's rhetoric responds to the limited conditions of poverty extant in the city in the period directly preceding the great immigration waves of 1890 through 1910. As immigration increased, and the economic growth and profits of New York's manufacturing and service economy came to increasingly rely upon the cheap labor market provided by the new arrivals, the search for a scientific solution to the urban problem grew more widespread and urgent. The period after 1890 witnessed an increased reliance upon the language of social science within both private and governmental committees.

The language of restructuring was also framed and contained by existing spatial conditions. Until the Tenement Law of 1901, regulations regarding tenement construction, and all attempts to devise new, more livable plans for the buildings themselves, were limited to those ideas that conformed to construction on the 25- by 100-foot lots created in 1811. Design competitions, such as one sponsored by the trade magazine *The Plumber and Sanitary Engineer* in 1878, produced little in the way of improved conditions, and are mostly remembered for producing the plans for the "dumbbell" tenements that proliferated from 1880 to 1901. Dumbbells were so called because the layout of adjacent buildings included a small, cut-out space between structures meant to conform to the law's requirement for light and air, with the images on blueprints resembling the handles and weights of the weight-lifting dumbbell. At the time of the competition, it was generally acknowledged that the restrictions imposed by the 25 by 100 lot made the construction of adequate housing profitable for investors nearly impossible. It was not until 1901, after numerous attempts to design successful dwellings, that architects and owners began to consider expanding lot size and opening up interior courtyards. Although there was no lack of advocates who suggested or demanded that profit take a back seat to livable conditions, the committees that directly influenced legislation, those appointed by and working within official government channels, were always balancing the needs of residents against the needs of accumulation.

An example of the contained language of committees and commissions concerned with restructuring can be seen in their positions regarding government sponsorship of housing. Although most housing experts and urban reformers such as Ernest Flagg, Felix Adler, and Phelps Stokes had been greatly influenced by European ideas regarding working-class housing, calls for direct government construction modeled on communities such as Huddersfield, England, and Duisberg, Germany, were rarely seriously considered. Resistance to state-financed public housing came not only from the private real estate interests that dominated New York's housing market but also from reformers like Riis, who believed that the market would eventually provide livable conditions for low-wage workers and their families. It was not until the severe housing shortage following the First World War that advocates came out openly in favor of public housing, and not until the Great Depression that any substantial housing would be built.

Decisions regarding spatial restructuring greatly influenced the daily rounds—lived space—of residents of the Middle West

Side. Although the area was represented by several powerful members of the Tammany Hall Club (including the infamous William Plunkitt) and was the subject of many studies by well-intentioned social workers, it lacked the financial and social power of more affluent areas. Thus no independent organization such as the West End Association existed to advocate for the needs of residents, at least until the founding of the McManus Club in 1898. As a result, decisions regarding the placement of parks, recreation piers, bathhouses, regulation of tenement construction, and removal of rail lines were largely made through the proceedings of committees and commissions that operated without input from local residents. These decisions greatly affected the daily routine of residents, determining where businesses would locate, what type of businesses they would be, how much open space would be available, and the general living conditions of the domestic interior. These decisions were taken within the confines of a language generated or constituted by existing spatial conditions. For residents of the Middle West Side, the decisions largely determined their lived space or daily rounds.

REPRESENTATIONAL SPACE AND PERFORMANCE
"No, dearie, I was never above Sixtieth Street in my life."

Included in the spatial archive of New York City is the monumental work by I. N. Phelps Stokes, *The Iconography of Manhattan Island*. This comprehensive, some might say obsessive work, published in 1926, provides not only a compelling snapshot of New York City between 1900 and 1910, but also a rigorous history of spatial change and restructuring in the city. Included in volume 6 of the eight-volume series are thirty-two plates forming a geographic archive of the city in or around 1910. Phelps Stokes's plates include topographical information, building identification, the location of rail lines, parks, baths, information of property ownership, and the location of sites of entertainment. The maps are an invaluable source for a reading of uneven geographic development in the city and a visual confirmation of the written reports compiled by social workers like Cartwright, True, and Anthony. Close study of the plates reveals both the striking diversity of the Middle West Side and the striking

Restructuring Progressives 75

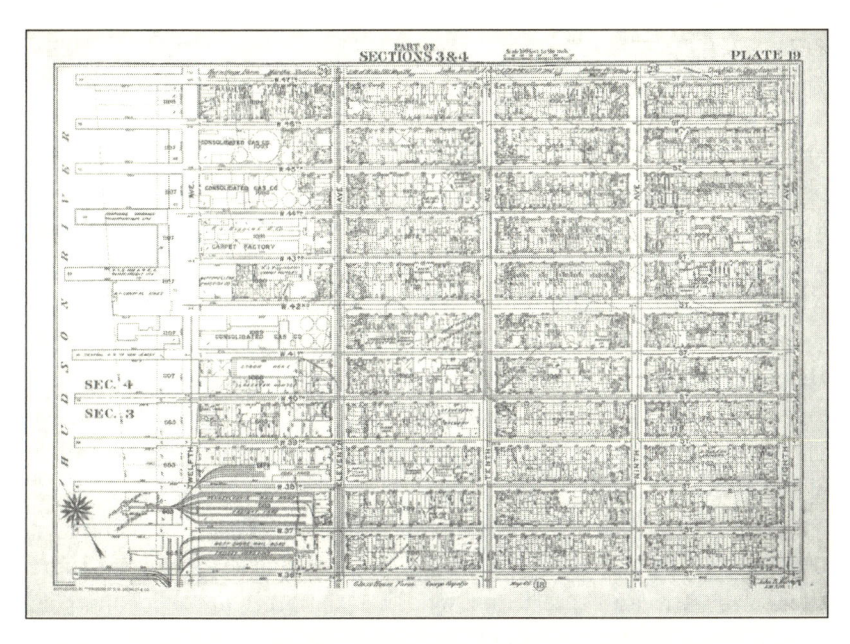

Phelps Stokes's obsessive mapping of the Middle West Side. (From *The Iconography of Manhattan Island 1498–1909*)

homogeneity of the Upper West Side just to the north. Plate 26, depicting the area between Central and Riverside Parks, shows a distinct lack of diversity in both building structure and use. Aside from the occasional school, church, hotel, and "named" apartment, the area is dominated by dwellings flanked on the west by Riverside Park and the east by the American Museum of Natural History, bordering Central Park. But Plates 18 and 19, depicting the heart of Hell's Kitchen and its environs, exhibit a stunning variety of building types, usages, and space, belying the descriptions of reformers who present the area as drab, monotonous, and devoid of character. Alongside and interspersed among the dwellings are breweries, factories, slaughterhouses, gas companies, stables, rail yards and freight lines, working piers, coal yards, settlement houses, model flats, dumbbell tenements, as well as hospitals, churches, social clubs, a recreation pier, and, from 1905, DeWitt Clinton Park, covering a full city block. The geographic contrasts reveal not only the excellent job performed by Cyrus Clark and other members of the West End Association in preserving the character of the Upper West

Side, but also the uneven development that culminated in the space of the Middle West Side, the Wild West of Hell's Kitchen, whose residents utilized this lived environment in performative actions based upon the representational spaces of their daily rounds. The performative nature of this particular brand of belonging was based in part on the perception of local spaces in relation to the wider city and nation, a neighborhood conception of citizenship often at odds with the dominant modes of belonging espoused by reformers.

For the first forty years of the nineteenth century, the area above Chelsea on the West Side was home to family farms and lots purchased by land speculators. The region, formally known as "Bloomingdale," was utilized as a summer retreat by the wealthy of lower New York, seeking escape form the ravages of cholera and dysentery that plagued the crowded regions of the city. German truck farmers purchased small plots, while squatters, many of them Irish and African Americans who had labored on the construction of the Croton Aqueduct, occupied the areas west of what is now Central Park. The construction of the park, along with the establishment of freight rail lines and working piers, increased the population from 1840 to 1870, bringing Irish and German laborers who worked the piers and toiled in the factories, producing iron, textiles, and lumber, among other products. One key event in the growth of population was the extension of the Ninth Avenue elevated passenger line above 30th Street in 1878. The extension of the line, along with increased surface service on horse-drawn streetcars, provided the region with the available labor pool for small manufacturing concerns, and resulted in the construction of rows of tenements on the 25- by 100-foot lots.[101] The combination of freight line access, proximity to working piers, and the availability of a flexible labor force made the Middle West Side an ideal location for a variety of manufacturing and service enterprises.

The period between 1870 and the "panic" of 1887 could be described as a golden age for the Middle West Side in terms of economic opportunity. During this period, the area was home to two large gas plants, Metropolitan Gas Company and Municipal Gas and Light, the rail yards of the Hudson River Railroad Company, Higgins Carpet, Travers Brothers Twine and Cordage, Stevenson's Brewery, Hardman Piano, and numerous smaller concerns producing finished wood products, textiles, glass, carriages, soaps, and dyes. The area was also home to print shops, iron foundries, mineral water and soda bottlers, and chemical manufacturers. But the space that probably did more than any other in defining the area, both for residents and outsiders, were the docks, the working piers that provided employment, but also provided numerous opportunities for crime.

The working piers of the West Side, privately owned and operated under city oversight, greatly contributed to the reputation of the area for many reasons. Charles Barnes, in his 1915 survey of dock work, was shocked by the unregulated nature of waterfront commerce. Barnes wondered why an industry so vital to the health of both the local community and the larger economy could operate with so little oversight, producing conditions of economic uncertainty, criminal activity, corruption, and dangerous working conditions. The piers of the West Side (including those in Chelsea above 14th Street) were a major source of employment for Middle West Side residents. Privately owned, the piers and the companies that controlled them were locked in fierce competition with waterfront operations on the East River in Manhattan, in Brooklyn, and on the Hudson in Hoboken, Jersey City, and Bayonne in New Jersey. The combination of stiff competition, lax city oversight and regulation, and proximity to spaces largely unfit for dwelling (as with the area along Twelfth and Thirteenth Avenues, parts of which disappeared under water during high tides) made the piers a defining feature of the area's reputation. Opportunities for theft of cargo meant that criminal gangs (like the Hell's Boys) often located near the docks, making the area well known to police and contributing to its reputation as "lawless." As well, the nature of longshore work contributed to the spatial reputation. Because loaders and unloaders only work when ships are in port, longshoremen are often idle, working long periods with equally long periods of unemployment, yet they must always be in the vicinity, as speed in unloading was a key part of the competition among stevedore companies. As such, longshore workers spent much of their idle time in the saloons that lined the dock area, sharing drinks and downtime with merchant marine workers from idle ships. Longshore work was dangerous, requiring skills in the proper handling of thousands of pounds of fast-moving material, which if not handled properly resulted in injury or death. Because of the piers' location near the underdeveloped waterfront, they were in close proximity to the slaughterhouses and stock pens in the areas of 39th to 42nd Streets, and 59th Street. These areas added greatly to the negative reputation of the Middle West Side, contributing foul odors and running waste products to an area already beset with smoke from steam engines on the freight lines. The relatively open spaces surrounding the stockyards and receding from the piers also provided space for ball playing and other activities of Middle West Side youths. Young people even used the polluted waters of the Hudson around the docks for swimming on hot summer days. These liminal spaces were also the site of scavenging, the collection of discarded (and often still-used) wood, lumps of stray coal, and other useful materials, which were collected mostly by children, often the younger ones who had not yet joined the official workforce.[102]

The intermittent nature of dock work mirrors the working conditions faced by most residents of the Middle West Side, particularly in times of economic downturn, such as 1887, 1893, and 1907. Although the area's access to docks and rail lines made it a prime location for manufacturing concerns, workers in the area were subject to the same pressures and insecurities attributed to our own era of globalization. As access to transportation infrastructure increased, companies prone to locate in the Middle West Side became increasingly flexible in their ability to move away from immediate sources of shipping, and locate near preferred labor markets. Work along the docks unloading imports was also affected by national politics when tariffs on foreign goods became one of the most contentious issues of the period. The economic history of the Middle West Side from 1880 to 1920 is one of constant economic dislocation and insecurity, as companies moved in and out of the area based on their relative ability to turn profits in the face of fierce competition. The relocation of the Higgins Carpet Company, something discussed in greater detail in chapter 5, is a prime example of spatial processes at work within an insecure economic outlook. Economic insecurity led many Middle West Side workers to look to organized labor as a source of protection. Indeed, the docks of the West Side were a hotbed of organizing for the Knights of Labor until their demise in the Panic of 1887. Employers, as in other areas, often responded to labor militancy by setting ethnic and racial groups against each other, as when the Ward Line Company broke an 1895 strike by hiring and retaining African American dockworkers, who had formally been excluded from such employment.[103]

While the piers of the Middle West Side waterfront served as a defining spatial feature of the area, particularly for outsiders, other spaces played important roles in the daily rounds of residents. The ubiquitous saloon served as the main source of social interaction for the adult male population, providing a space of sociality and also a meeting space for labor and ethnic organizations. For young girls and boys, basements often served as informal meeting spaces for clubs that were almost exclusively divided by gender. Settlement houses also provided space away from the tenement interior, but these were regarded with suspicion by many residents, who viewed settlement house workers as meddling outsiders, and often they were used reluctantly, only as a last resort. Tenement rooftops were a main space of social interaction, providing room for supervised courtship on summer nights, as well as for kite flying and pigeon keeping. Church halls also provided social space for area residents and were favored by women as safe spaces for social interaction.

In the routine paths from spaces of employment to spaces of dwelling and sociability, Middle West Side residents made their own space, but not under geographic conditions of their own choosing. Like reformers, politicians, and landlords, they constructed a common language of understanding within which they could perform their own versions of belonging and citizenship based on their perception and interpretation of the environment— the representational space of the everyday. How these residents read their spatial environment, in both the sense of negotiating the daily rounds and interpreting the spatial environment, can be best explained utilizing Kevin Lynch's categories of paths, nodes, districts, edges, and landmarks. For Lynch, each of these categories serves as a marker or signifier of readability, serving to orient urban residents to their surroundings. Paths are defined as areas of transport, nodes as "strategic foci" or points of concentration, such as markets or rail terminals. Edges serve as linear boundaries that define and demarcate districts. Landmarks serve as physical reference points that orient perception. As Lynch points out, none of these elements can exist in isolation, and the same spaces or locales can serve as orientation points in several of the categories.

"Districts are structured with nodes, defined by edges, penetrated by paths and sprinkled with landmarks. Elements regularly overlap and pierce one another." It was these categories that were the focus of reform efforts to provide visibility to the Middle West Side, through attempts to restructure the daily paths, nodes, and edges, and provide proper living space.[104] Below, I will briefly describe how different residents of the Middle West Side negotiated their daily rounds in the liminal space between visibility and occlusion. Subsequent chapters will detail how spatial restructuring affected daily rounds, producing in effect the spaces of production that generated languages of difference, desire, insecurity, and belonging.

The daily round for Middle West Side residents varied greatly, depending on age, occupation, gender, social situation, education, and other factors. But a composite picture can be drawn from the historical record based on a commonality of experiences related to these categories. Further, the representational space of residents, how they perceived and interpreted their environment, also varied greatly, but common experience can provide telling detail. For women of the Middle West Side, the common understanding is of the urban provincial, as represented by the quote at the beginning of this section, "No, dearie, I was never above Sixtieth Street in my life." Though this reality was typical of many women tied to both domestic labor and employment in wage or piecework in the MWS, the experience of daily life was diverse, dependent on the circumstances listed above. Middle West Side women's experience of daily life ranged from the limited horizon of the typical urban provincial to the "bohemian" existence of single women who occupied the Windmere model tenement at 57th Street. But by far the most common everyday experience for women on the Middle West Side was played out within the confines of the extended family unit typical of the period and area.

For most women, the dividing lines of life experience were determined by stages within the hetero-normative lifeworld: childhood, young adult, and married woman. This bracketing is particularly to the historical period and the urban scale, as mandatory schooling and child labor restrictions separated childhood from "youth," the period after elementary education when young women were sought by certain employers based on their willingness to accept low wages. The bracketing is continued after marriage and childbirth, when women's work frequently included the running of the household, taking in work such as laundry or sewing, and outwork at nearby businesses, often as cleaners in theaters or office buildings, or as housekeepers. Income disparities and occupational differences also were determinative of daily rounds. The wife of a merchant with a shop on Eighth Avenue and dwelling above would have a different daily experience and outlook than that of a working mother who lived on Tenth Avenue. Teenage and twenty-something factory girls traveled a particular route, mentally and physically, than that available to the mother of four in her thirties. Young single women involved in entertainment, such as those who worked in theaters, experienced more aspects of the city, had different paths and landmarks, than that of the merchant's wife tied to the store. In fact, single women often worked seasonally, working in factories during peak seasons and as cleaners in downtimes. One Middle West Side woman reported that she worked as a cleaner in the West Side theaters in the winter, and migrated to Atlantic City in New Jersey for summer work as a waitress.[105] This type of varied experience was usually reserved for single women, as married women's horizons were restricted by the need to return home. And for children under working age, daily experience was often limited to the immediate vicinity of the tenement, school, and surrounding streets.

The daily path and hours of work varied according to age, income, and occupation. Even with the availability of street trolleys, an elevated train on Ninth Avenue, and cross-town trolleys, most women worked locally due to the prohibitive price of fares compared to their daily earnings, and for the need to return home. The majority of working women were classified as domestic or personal service,[106] and many worked in theaters in the West Forties, homes on the Upper West Side, or office buildings in the mid-Fifties. Walking was often the mode of travel. To walk from a tenement apartment on Tenth Avenue and 38th Street to a job cleaning Lew Field's Music Hall, a vaudeville house on 50th and Seventh Avenue, often in the early morning or late evening, meant negotiating a path that took one near the dives of the Tenderloin, several breweries, and numerous local taverns. The return trip could include a stop at "Paddy's Market," on Ninth Avenue between 45th and 40th, under the elevated.[107] Shopping at the pushcarts that lined the darkened street, one could expect the noise of the overhead train, perhaps a hot ember falling on your shoulder from the steam engine, the smoke billowing, and crowds trying to strike the best deal with the vendors. Returning home, one would be close enough to hear the rumble of freight trains at their yards between Eleventh and Twelfth Avenues, and to smell the slaughterhouse and freight yards two blocks north. For the West Sider, locations such as Paddy's Market served as nodes, focal points where socializing could occur and information shared. Large businesses such as breweries, with their distinctive smells, docks open to the streets, and buildings that occupied one-third of a city block, could serve as landmarks, orienting points. For a woman walking alone at night from a job at Lew Field's, cutting over to Ninth Avenue and proceeding south, avoiding the Tenderloin district and traveling an "edge" might save her from potential harassment, or even arrest on suspicion of prostitution.

Young, unmarried women, statistically more likely to work in manufacturing, would often recognize different nodal points. For such women, the place of employment, where much of the day was spent and large groups spent long hours at routine tasks, would serve as a space of sociality. These young, unmarried women were also likely to frequent the dance halls of the theater district and Tenderloin, and were more likely to spend part of their income on public transportation, forging paths into the larger city that were closed to women with familial obligations. For girls not of working age, the daily path often included the early morning "scavenging" run that was expected of school-age children, which could include lining up for charity milk and cheese distribution, gathering coal and wood, and buying ice for the preservation of food.

For many men, the daily path led from home to work at the waterfront. But many also worked in transport, traveling some distances to reach their place of employment. For many, social life revolved around the saloons that lined the avenues. For all residents, the streets, sidewalks, and rooftops were places of socialization, where norms of behavior were enforced and contested. Men and boys engaged in pigeon keeping and kite flying off the tenement rooftops. Boys often played at the waterfront, but streets also provided places of play, and using them risked a visit from the beat cop.

Restructuring had different effects at different scales for Middle West Side residents. Expansion of public transportation allowed

A knife sharpener scrapes out a living, literally, on Ninth Ave. and 54th St. (Milstein Division of United States History, Local history & Genealogy, The New York Public Library, Astor, Lenox and Tilden Foundations)

locals to expand the areas of available employment. It also meant that workers were less tied to their area of employment, encouraging many residents to seek better housing opportunities elsewhere. Improvements in housing meant better living conditions for some, while others were caught in substandard housing. Park, bath, and pier construction created spaces of play for young people, allowing them to create spaces of sociality. The establishment of settlement houses provided spaces of "healthy" activity, but also created the impression of a paternalism that many residents rejected. As a result of a long history of neglect from city authorities, many efforts to "improve" conditions in areas like the Middle West Side were met with skepticism by residents. Some residents utilized the move to reform and progress to demand further improvements in their neighborhoods, while others took advantage of the new opportunities for mobility to abandon the area altogether.[108]

The lived space of the Middle West Side provided the structural setting for the performance of a variety of approaches to daily life, and the restructuring of that space contributed to changes in attitudes and perceptions among the local residents. The paths, nodes, edges, and landmarks of daily life were altered not only by the physical restructuring brought about by city planners and reformers, but by their daily use. The construction of DeWitt Clinton Park in 1905, occupying two complete blocks, would create a new landmark, a node of gathering, and possibly a boundary or "edge" for some residents. For this and smaller changes, like the installation of a bathhouse or establishment of a settlement house, the use of such structures, whether they became nodes or landmarks, depended upon the attitudes of residents toward them. Expansion of public transportation lines, such as an east–west streetcar, opened up areas of the city, but only for residents who could afford, monetarily and temporally, to utilize them. Street cleaning and road repair made the area more attractive to businesses, supplying increased employment opportunities, and improvements in transport and communication enabled businesses to relocate out of the area, causing economic distress for local workers.

As the following chapters will demonstrate, Middle West Side residents developed norms of behavior and patterns of action based upon both their own experience of the local space and their awareness of and encounter with the conceptual ideas of city authorities and urban reformers. As the area became a subject of study for reformers and remained a zone of criminality for authorities, residents often internalized these views, constructing identities that reflected the spatial understandings of reformers, authorities, and themselves. Performing these spatial identities meant, at various points, resisting the efforts of social workers, participating in black market activities, defending neighborhood "turf," forming and reforming ethnic, racial, and religious allegiances, and testing the boundaries of gender and sexuality. The "wants" of Middle West Side residents were expressed through these spatialized identities, and the spatial environment, the area where they lived and worked, constituted or framed the demands they made upon city authorities. As anthropologist Partha Chaterjee suggests when discussing the demands of contemporary South Asian slum dwellers, these demands are often different from the claims of "citizens," a particular identity increasingly dominated by an expanding middle class and affirmed through the mediation of both the nation and the capitalist economic system. The claims were more often advanced through participation in illegal activities and by turning their strategies of survival into "legitimate knowledge about surviving poverty."[109] By participating in the production of space, residents of the Middle West Side forged their own norms and understandings regarding citizenship and belonging, but they did this under spatial conditions not of their own choosing.

CHAPTER THREE

WHEN HELL FROZE OVER

"We used to fight the Irish kids. They used to call us guineas all the time, or wops, or meatballs. They used to say they were better Americans than us. But we were born here too."

"The Tenths and Elevenths (avenues) would come down to fight. When they started, we would try to get our brothers and mothers and fathers. Our parents fought for their children. I remember Grandma going after them once. She saw them grab one of my cousins and she whacked them. Mostly, though, there was a lot of name-calling."

"These poor unfortunates they arrested for fighting with their wives or stealing watermelon were taken down to the 37th Street station house. We used to climb up on the iron rail outside the building to see what was happening inside. 'Oh, he just belted him again.'" [110]

ON JULY 4, 1911, what newspapers reported as a "race riot" took place on Eleventh Avenue and West 40th Street. Although the brawl occurred during a period of widespread racial violence in many areas of the United States, this particular fight involved not "whites" attacking African Americans, but rather what the *New York Times* described as "old-time residents" attacking Austrian immigrants. The details reported by the *Times* are striking for the headline that declares the brawl a race riot, yet only one ethnic group is mentioned by name in the description, a small group of "Austrians" that the report claimed had recently moved to the area. The opposing side in the so-called race riot is described as the Hell's Kitchen gang, as in "old-time residents of the district had watched with increasing anger the encroachment of Austrian immigrants into the area."[111] The unspoken "race" is meant to signify the West Side Irish, whose presence had become a defining characteristic of the area. The newcomers, based on their reported surnames and on immigration numbers, were most likely Croats. The use of the term *race* is typical for a period when social scientists, politicians, and public figures commonly divided even "white" Europe into such categories as Alpine, Nordic, and Slavonic, with each group retaining distinctive characteristics linked to geographic origin.

In the incident, a fight between rival "Austrians," possibly over an unpaid debt and name-calling, had escalated into a "race war," as the "old-timers" used the incident as an excuse to destroy the tavern of one Anton Ruscovitch, a social gathering spot for the "newcomers." The story reports that though the "gangs had been subdued" in Hell's Kitchen, and policemen felt safe enough to walk beats alone, the old neighborhood still had some "spirit" left in it, enough to rouse the local population to violence, taking a vaguely defined "vengeance" against the Eastern Europeans for making inroads into their neighborhood. The attacking Irish allegedly used paving stones to break through the plate-glass windows of the bar, and proceeded to wreck the interior, as "tables and chairs were broken and every bottle and mirror in Ruscovich's saloon was smashed." As the police arrived, the attackers mysteriously faded back into the neighborhood, and the police found "none left to arrest." The violent incident fit neatly into the stream of reporting about the Middle West Side during the period, when reporters, journalists, and novelists treated violence, and particularly "race" violence, as part of the everyday fabric of life in the Middle West Side.

New York City had witnessed its share of riots, with Hell's Kitchen being the primary stage for perhaps the worst outbreak of racial violence, the 1863 Draft Riots. Large-scale public violence is recorded in the Middle West Side in 1898, 1900, 1901, and 1911, and racial divides were usually the central issue in these incidents. Questions of race and racial construction had played important roles in city politics, identity construction, and certainly in issues

A familiar sight for Hell's Kitchen residents: a street cleaner and police officer. Both departments, Police and Sanitation, were objects of reform. (New York City Municipal Archive)

surrounding uneven geographic development. But though the intersection of race, ethnicity, and public violence is certainly important in the production of space on the Middle West Side, I am concerned here with a different historical trajectory, one intimately embedded in racial tension and inclusionary/exclusionary violence. Although not denying the power of ethnic, religious, racial, and class identities to produce feelings of propriety, solidarity, and protectiveness, and produce, among urban populations, what Chicago urban sociological pioneer Robert Park famously described as "a mosaic of little worlds," my focus is on how certain types of difference are produced geographically under contingent historical conditions and relations of power, and how such conditions produce the spaces in which violence and other forms of contestation are performed.

In this chapter I examine the intersection and interactions of public authority, mainly in the form of the New York City Police Department, and demonstrate the often subtle but sometimes obvious ways in which the actions and attitudes of authorities created the conditions for what I term the production of spatial difference.

I will show how the lived space of the Middle West Side produced a discourse of criminality, disorder, and occlusion that permeated both public perceptions of the area and the actions of city officials. The reality and perception of criminality led city officials to attempt to solidify the porous boundaries of the Middle West Side through the creation of "frozen zones," areas in which either known criminals are arrested on site, or where crime is contained through a variety of both officially condoned practices and illegal ones, such as the collecting of bribes and graft. It was the particular spatial conditions created by these police zones that contributed to the formation of shifting identity groups within these areas, and thus the production of spatial difference.

Difference here refers to the production of identities that lay somewhere between Park's "mosaic" of ethnic enclaves and a Derridean realm of pure play of "inequalities that make possible the movements of life." Through the creation of "frozen zones" of criminal activity and police corruption, city authorities created spaces in which law officers acted with impunity, denying the basic citizenship rights of the residents. By attempting to solidify the boundaries of areas of perceived criminality, police officers and other officials caused shifts in other boundaries within communities, as ethnic, racial, class, and religious groups formed and re-formed in both reaction to these conditions and as agents in their construction. Although these formations and re-formations sometimes led to interethnic coalitions in opposition to established authorities, they more often led to violent clashes among identity groups who were all engaged in the struggle to secure their rights to full citizenship.

The difference produced by the construction of frozen zones is largely the result of the uncertainty created through the deployment of arbitrary power. As industrial geographer Doreen Massey has suggested, the construction of spatial zones—regions, neighborhoods, individuals—is a process of interaction and engagement "forged in and through relations (which include non-relations, absences, hiatuses)."[112] The "relational nature of space" produced through the interaction and engagement between Middle West Side residents and public authorities breaks down the obvious boundaries of race, ethnicity, gender, and class by focusing the actions of residents on the struggle to obtain conditions of security and geographic citizenship against the arbitrary power deployed against them. Frozen zones, like other instances of arbitrary power within a system of law, create conditions of uncertainty in which individuals and groups try to reconcile their positions as official "citizens" with full legal rights with their treatment by those exercising direct forms of legal authority. This uncertain status works to re-solidify certain forms of identity while simultaneously breaking them down by demonstrating the porosity of cognitive, internal boundaries, as impunity is exercised against residents across identity lines. Thus this breaks down the "mosaic" of little worlds into worlds that "interpenetrate, and at least overlap, as well as touch." Interrogating the ways in which the deployment of arbitrary power structured the formation and re-formation of spaces of identity can, as urban planner Kian Tajbakhsh suggests, illuminate how "boundaries and borders are both productive and subversive of identity, rather than simply lines of demarcation between mutually exclusive groups," and the areas of engagement and interaction that produce difference.

Most urban residents and visitors recognize boundaries, the distinct or near-invisible markers, often spoken of in barely coded terms, beyond which there is danger, usually in the form of black market activity. Urban boundaries are sometimes obvious: they can literally be train tracks with a "wrong side." They are often less visible, disputed, crossed, drawn and redrawn, fought over, restructured, and can be based on a variety of factors, including class, race, ethnicity, zoning, access to public transportation, provision of municipal services, and proximity to industry, rivers, or open space. Certain boundaries are official borders, codified in municipal maps and demarcated by signs and directional aids. The most important, and probably most permeable of boundaries, are the cognitive borders shared among urban residents and visitors.

The boundaries of the Middle West Side, the porous, crossable zones that define this or any urban region, seem to make the moniker Hell's Kitchen almost inevitable, an occurrence of God and nature, like the rising property values of the Riverside Drive area, as discussed in chapter 1. This seeming inevitability is evident in the tone of many of the social surveys written about the area, in which the authors express a pained sympathy for people whose place of residence is bounded on the east and west by two areas that are the breeding ground for criminality and unofficial economies. On the west lay the Hudson River and the docks that serviced the ships arriving from domestic and international ports. Shipping piers and loading docks are prime locations for crime, as numerous opportunities for theft exist among the incoming and outgoing freight. Unregulated docks traditionally have drawn members of organized crime units, who attempt to get a piece of the theft action while also controlling the labor force through violence and intimidation. The western boundary was both physically and cognitively definitive for the Middle West Side, as the river drew a strict geographic line and the port activity defined for both residents and outsiders alike the rough-and-tumble reputation of the area.

To the southeast, the Middle West Side was bounded by the notorious Tenderloin district, a boundary more porous than its western counterpart. The Tenderloin was a well-known location of saloons, dance halls, and prostitution, serving a diverse costumer base that included locals, downtown businessmen, tourists, and the occasional crusading reformer and investigative journalist. The district blended into the theater and entertainment area to the north, and was bounded on its own eastern edge by the Broadway shopping areas, the redoubt of the respectable middle-class woman, kept clear of vice by the New York Police Department's "Broadway Squad," who contained illicit activity within the Tenderloin proper. The theater district, home to playhouses hosting serious productions and to vaudeville theaters, with their mix of mid-level and low brow folly, provided residents of the Middle West Side with both a place to seek affordable entertainment and a place of employment—as ticket-takers, cleaners, ushers, and sometimes, performers.

To the south lay upper Chelsea, itself a blend of factories and tenements, though without the reputation of the Kitchen. The northern edge of Chelsea blended easily into the Middle West Side, and its southern edge was a largely middle-class residential area, containing among other landmarks Madison Square Park and the First Presbyterian Church, home of the crusading reformer Reverend Charles Parkhurst and his Society for the Prevention of Crime. The northern edge of the Middle West Side was both more distinct and ambiguous at once. To the river, starting in the upper Sixties, was the Upper West Side, a zone of bourgeois housing bounded in the Seventies by Riverside Park and maintained as residential by the powerful West End Association. But the lower Sixties and upper Fifties contained the area that came to be known, after 1898, as San

San Juan Hill, Hell's Kitchen's African American district, as it looked in the 1940's. (La Guardia Wagner Archive)

Juan Hill, the home to many of New York's African Americans and immigrants from the West Indies. The area had garnered its nickname in 1898 after the Spanish-American War as the New York site of battles between African American and Irish groups. Many African American families had settled in the area after fleeing uptown in the wake of the Draft Riots of 1863. The clashes between "whites" and African Americans in San Juan Hill and other areas in the region were often the result of shifting cognitive and physical boundaries, and the result of the pressures of living within zones that had been designated by public authorities as areas of high crime and vice.[113]

Like all regions, the Middle West Side contained its own system of edges and boundaries, some definitive and lasting, others shifting and in constant dispute. Class, income, and occupation formed some of the internal boundaries. Moving west from Eighth to Eleventh Avenue meant passing from neighborhoods occupied by merchants and skilled tradesmen on Eighth, with higher rents and better housing stock, to the "Avenue of Death" on Eleventh, where the cheapest tenement rents were found. Eleventh Avenue, with a freight train line running down the middle, housed some of the area's poorest and most itinerant residents, and was considered by both insiders and public authorities as the "roughest" area of the district. Rents were lower along Eleventh Avenue due to its older housing stock, the freight line, and its proximity to the piers and slaughterhouses. Divisions existed among members of the same religion based on class and income markers, as the better-off Irish attended St. Malachy's on 47th Street, and the less affluent attended the Church of the Holy Cross on 40th. Holy Cross distributed charity to the destitute among the Middle West Side Irish, something St. Malachy's parishioners looked upon with thinly veiled contempt. Who would or would not accept private charity was another internal dividing line.

Although there is evidence of ethnic clustering in certain streets and tenement buildings, and evidence of ethnic solidarity in times of violent disputation, divisions of "turf," particularly among the young, appear more geographically than ethnically based. Based on survey data, most tenement buildings seemed to be ethnically mixed, with only African Americans consciously excluded, but even this exclusion was not complete. Although the Irish were the dominant ethnic group in the area, they intermarried often, and their ethnic solidarity seems largely rhetorical, as in the use of pejorative monikers for other ethnic groups, and a general sense of ethnic superiority. As in most neighborhoods, first loyalty was to immediate family, and secondary solidarity seemed to lie mainly with the street one lived on, especially among adolescent boys and girls. From the accounts gathered by social workers, play and socializing was interethnic, a situation that was maintained often until a serious dispute arose, with different ethnic groups retreating into ethnic solidarity only if a larger dispute forced the breakdown of geographic social circles. The rule seemed to be: you can be friends with the Irish, Italian, Slav, but you can never really trust them.

Internal edges also included gendered and sexualized boundaries, enforced through informal systems of normative behavior. Most saloons were the domain of men, with separate "family" entrances where women, or children, could enter to purchase cardboard "buckets" of beer to be taken home. The tenement interior, including hallways, was the domain of women and children, places of socialization and interaction that were largely devoid of male presence for most periods. Cellars and their entranceways were considered danger zones for children and adolescents. Although statistical evidence indicates that sexual assault was as likely to occur in the home, most children were told to avoid going into basements and to report incidents of adults or older children trying to lure them there.[114]

The boundaries themselves were physical and cognitive, solid and permeable. Physical boundaries such as the river and its piers and the western edge of Central Park were easily recognized, and there was general consensus as to their meaning and orienting power. But physical edges were cognitive at the same time, constituting the informal boundaries between neighborhoods that most residents recognized in their daily routines. Middle West Side residents could surely recognize the difference in the housing stock of their own region and the Upper West Side, and were surely aware of the "mixed use" makeup of their neighborhoods as compared to the residential order of the Upper West Side. Middle West Side residents and visitors were consciously aware of class markers such as dress and deportment, as witnessed by the numerous accounts of "respectable" citizens receiving "rough" treatment from neighborhood youths simply for entering the area. This worked in the other direction as well. A news account reports women shoppers from affluent neighborhoods who ventured into Paddy's Market seeking medicinal herbs, "leaving their carriages at a convenient distance," and in spite of these precautions "are at once spotted."[115] Middle West Side residents were aware as well of their position as the "object" of study of the reforming classes and often played the role to their advantage. Though the physical edges that framed the area were the river and numbered streets, Middle West Side residents were also defined by the perception of the neighborhoods in the minds of outsiders, like reformers and city authorities. Their view of themselves as citizens, members of a larger community defined by both law and tradition, was greatly affected by geographic realities and the geographic understanding of those in positions of power and authority. In the "relational construction of the identity of place," no relationship was more important than that between the residents of the Middle West Side and the unique carriers of state power and authority, the police, whose role as enforcers of public order involves the maintenance of public space and the solidification of permeable boundaries.

German philosopher Walter Benjamin, in *Critique of Violence*, writes of the "ignominy" of police, particularly when operating under modern conditions of the "rule of law." For Benjamin, the public shame of policing stems from the unique position, or "unnatural combination," of police officers, who must act as both the preservers of the rule of law while preserving public order, often at the point of dispute where the proscriptions of law are often vague and cannot govern every situation. Police officers at the point of contact with the public thus become both law-preservers and law-makers, a situation that, for Benjamin, allows police to perpetrate "the crudest acts" and to "rampage all the more blindly in the most vulnerable areas." What Benjamin is describing is the condition of impunity, where police activities clearly violate the legal rights or basic dignity of the individual, with little or no fear of punishment or sanction. Police impunity can take a variety of forms, particularly when police operate under the restrictive conditions of positive law and constitutional protection. As Benjamin points out, police violence and impunity is "less devastating" when they represent regimes that overtly deny basic rights. Operating as the preservers of public order and public space, police under conditions where rights exist to protect citizens against arbitrary action have developed established repertoires for enforcing order. These include the use of such vague statutes as "disorderly conduct," "disobeying a police order," "disturbing the peace," and the enforcement of laws concerning vagrancy and orderly public space.

The application of public order statutes at the point of contact with the public allows police officers the level of flexibility and discretion necessary to "legally" violate the civil rights of citizens. Vaguely defined public order statutes such as "disorderly conduct" can be applicable to an entire range of activities, and are open to police interpretation. This lack of effective and consistent codification has often allowed police officers a cover for other motives for arrests while maintaining a veneer of legality. Police in areas like the Middle West Side would often arrest those who worked for opposing political camps, refused to pay graft, filed charges against officers, or against whom they simply did not like. Public order statutes could also be utilized during "sweeps," those periods when for various reasons, the local precinct was ordered to "clear" the district, and police officers would arrest "known" criminals on "suspicion" of criminal intent.

Further complicating the situation for Middle West Side residents at the turn of the century was the nature of civil rights recognition by police officers and civil courts. The "right" of the citizen against arbitrary state action, though enumerated in the Bill of Rights, has also been subject to interpretation, as different courts in different periods have recognized a variety of applications of basic civil rights. This is particularly true in cases regarding the right of police to maintain "order" in public space, where courts have only recognized the "rights" of citizens to immunity from police harassment over a long span of evolving legal theory.[116] Further, as with most urban police forces, the NYPD had developed an internal code of behavior in which officers supported each other in testimony, often with the active aid of judges, manipulating the evidence and narrative of events. When these factors are combined with the tendency of judges and juries to accept or reject testimony based on the witnesses' class, ethnic or gender identity, it leads to situations where police officers feel unconstrained in acting to violate basic rights. As Marilynn Johnson points out, police officers are trained in "the proper articulation of probable cause. . . . suggesting objective circumstances of a police action [that] could be reconstructed to fit legal requirements."[117]

But for residents, the most important spatialized aspect of this relationship with public authorities was their perception of the Middle West Side district and residents' own perception of the role of those authorities in respecting their rights as full citizens. By the 1850s, the Middle West Side had garnered its reputation for lawlessness. It had also, by the 1870s, developed as a district of low-rent properties, where working-class families and individuals could afford to live. For many of these families, as with most families in any "slum" district, the daily routine consisted of seeking work, providing for self and family, and looking for sources of sociality and entertainment while seeking a measure of security. Unfortunately for Middle West Side residents, they sought these basic human needs in a district that had become "frozen," an area of crime containment where officers of the state attempted to solidify permeable edges and were allowed to act with impunity. The freezing of the Middle West Side, the attempt to solidify the permeable boundaries or edges, and the consistent violation of basic notions of citizenship led many residents to develop a negative, often confrontational attitude toward not only the police but to city authorities in general. To understand how the freezing of Hell's Kitchen came about, it is necessary to briefly review the history of policing in the city, and to look at the ways in which authorities made both conscious and unconscious decisions to violate the basic rights of residents.

Just as New York City can claim to have established the first true municipal police force in the United States, it can also lay claim to having the first government-ordered investigation of police corruption. The two events are not unrelated. New York City's police force was established by an act of the New York State Legislature in 1845 in response to the fears of "respectable" citizens of the growing "slum" districts in the Five Points, Bowery, and East River areas. The police force was established to protect the property rights of middle- and upper-class Manhattanites against the "dangerous" classes and to maintain public order. Though it was claimed that the NYPD was modeled on the London Metropolitan force, control over the force, unlike in London, was left to "local" control, meaning the appointment of officers and administrators was left to the discretion of the mayor's office and local ward leaders. Mayor Fernando Wood, credited with developing some of the important internal parts of urban machine politics, seized the opportunity to politicize the force, allowing local ward leaders to dispense police jobs in exchange for services, such as "help" on election days. Wood's actions led the state legislature to create its own police force, and they promptly attempted to arrest Wood for his dismissal of a state-appointed sanitation chief. As a result, New York City in the financial panic summer of 1857 was "subject to the spectacle of rival police forces patrolling the streets and vying for custody of police buildings."[118] Though the state won the battle in 1857, by 1870 it had lost the war, and control of the NYPD returned to the city as part of the Tweed Charter. Under the control of various administrations, the NYPD became a political football, and day-to-day policing was often a matter of what political faction was in charge and whether they needed to use the police to promote their supporters in the vice businesses or prove their credentials as reformers.

Several dynamics combined to create the conditions for spatially contained police impunity. First, control of police appointments meant a constant revenue stream for the party in control, as well as the guarantee of Election Day aid. Ward leaders often appointed loyalists to the force, who would use their position to collect graft from illegal brothels to pad their salaries. Before 1900, police officers often purchased their appointments, and made up the purchase price by squeezing local businesses, both legitimate and black market. For even legitimate businesses, the typical arrangement consisted of paying a set amount for an official excise license, and paying extra for police protection.[119] Though the salary of the patrol officer usually maintained itself at twice the average pay of an industrial laborer, the opportunities for extra money, through the taking of graft, enhanced the desirability of the job. Positions on the police force were often paid for directly to the ward boss, particularly promotions to precinct captain, a lucrative position, especially in areas like the Tenderloin. Police made their money by accepting payments either from illegal establishments, such as brothels, or from legal establishments that conducted illicit activity, such as gambling, prostitution, or Sunday alcohol service. Most appointed officers would vote for the faction that gave them their jobs, but their real service at elections was policing the polling places, which often meant ignoring the strong-arm intimidation tactics of representatives of their particular faction.

Second, because the force was politically controlled, it was often utilized to placate certain segments of New York society depending on political expediency. For instance, if merchants on the edges of the Middle West Side along Eighth Avenue threatened to withhold support because of rising crime, the area would be "swept" by strong-arm squads, groups of nightstick-wielding police who would indiscriminately clear the area of all known criminals and suspicious persons. Conversely, ward bosses and mayoral administrations depended on the support of both the owners and customers of saloons, brothels, and gambling dens. "Crackdowns" were often made for the benefit of the press and moral reformers, while zones of vice were maintained by containing them within certain areas such as the Tenderloin, San Juan Hill, and Hell's Kitchen. Finally, the use of the police as a political tool meant that individual officers who acted on behalf of the ruling faction were seldom subject to disciplinary procedures, at least until after 1900. This type of impunity was not restricted to police who supported Tammany or any machine. Zones of vice were an accepted part of the city economy by all political factions. The major difference lay in the fact that reform, Republican, and fusion administrations usually placed increased emphasis on "cracking down" on vice to please their supporters, a situation that probably increased the incidences of police impunity against ordinary citizens. Even reformers such as future president Theodore Roosevelt, while serving as head of the police commission, objected only to violent impunity against "respectable" citizens, and considered the indiscriminate use of the club against the "criminal element" to be a manly approach to the problems of street crime and vice.[120]

The politicization of the police force and the economic and social realities of the Middle West Side, combined with the fears of the propertied classes of the working classes, worked to produce the space that encouraged and maintained police impunity. In the 1901 Tenderloin Riot, discussed later in this chapter, how authorities judged an individual citizen in public encounters was embedded in the spatial context. Residents of areas of contained criminality were also at risk from police officers for two additional reasons. First, areas like the Middle West Side became dumping grounds for officers accused of violations of police regulations. The area was often patrolled and commanded by the most violent and lawless officers on the force. Second, because the MWS did not contain the opportunities for graft that areas like the Tenderloin did, officers looking to enrich themselves often squeezed money out of sources that were barely able to afford it, unlike the lucrative brothels and gambling dens to the south and west. The take for a single, high-end brothel equaled that which could be squeezed from dozens of low-end saloons and unlicensed pushcarts.

As a result of the widespread abuse and political use of the police department, the New York State Legislature, under pressure from moral reformers led by Reverend Charles Parkhurst and the Society for the Prevention of Crime, created the Lexow Committee in 1894 to investigate police corruption.[121] The investigation, producing over ten thousand pages of evidence, "uncovered" many things that most New Yorkers already knew regarding the NYPD. But the testimony taken by the committee was enough to cause a reaction against the more widespread abuses. Aside from the revelations concerning bribery, graft, patronage, and political favoritism, the committee testimony by police officers showed the explicit and implicit nature of urban policing and the connection between class, ethnicity, space, and perception that worked to produce the frozen zones of police impunity. Two issues that were not at the top of the priority list for reformers like Parkhurst concerned policing and spatial containment: the use of the official "frozen zone" and the creation of unofficial zones of criminality, and the indiscriminate use of force used predominantly against citizens in certain areas.[122]

The contemporary use of the term *frozen zone* in relation to policing usually refers to efforts by officers to create zones in which certain activities are allowed and zones where they are restricted. This is especially true in the practices of public protest policing, in which authorities set up demarcated protest or "free speech" zones, and frozen zones around which no one is allowed to enter during major political events, such as World Trade Organization (WTO) meetings.[123] This practice of spatialized policing has a long history, particularly among urban police forces. The creation of urban forces like the NYPD was brought about, in part, in order to protect the property rights of merchants, landlords, and homeowners. New York City's first experience with police zoning was probably the creation of the "Broadway Squad," made up of specially selected officers of intimidating size to assist female shoppers in crossing the busy boulevard, but whose main purpose seems to have been to keep the street safe for middle- and upper-class consumption. Under the guidance of Thomas Byrnes, the NYPD detective squad established in the 1880s a frozen zone of exclusion, in the Wall Street area. The officially declared police boundary, extending north to Fulton Street, demarcated an area in which criminal suspects, those known to the police, or even "suspicious persons," were barred from entering for any purpose. Byrnes also established special detective squads to work solely on the recovery of property stolen from wealthy New Yorkers. The official zone, combined with unofficial emphasis on containment, created areas of exclusion that, in effect, confined actual criminal activity within the boundaries of other areas, which became part of the accepted practice of policing, and exacerbated the already existing crime problems located there.

For the Middle West Side and its residents, vice and the creation of frozen zones that were "cleared" of criminals and suspicious persons exacerbated the physical and social isolation of their area. Historian Timothy Gilfoyle's study of prostitution in the city shows the Middle West Side from 1880 to 1910 as an area almost completely surrounded by brothels and houses of prostitution.[124] The Tenderloin to the southeast was heavily populated with gambling dens, bawdy houses, and saloons where criminal activity was protected. As Gilfoyle and others have demonstrated, the criminalization of vice in the Tenderloin and the subsequent "licensing" of places of vice activity was a lucrative source of income for police officers, wardmen, captains, and the local political club. Although prostitution was commonplace in various parts of the city, its greatest concentration by 1900 was in the Tenderloin. Though its famous bawdy houses and theaters attracted crowds from all the city's social classes, the Middle West Side did not develop as an "entertainment" district for the city's "sporting men." Comparatively, the Middle West Side west of Eighth Avenue contained relatively few brothels. But proximity to the Tenderloin meant that during the city's periodic crackdowns on organized vice, prostitutes forced out of Tenderloin brothels often took up residence in the tenements of the Middle West Side. This fact, combined with the tendency of a small number of working-class women to seek higher wages through "casual" prostitution, contributed to the area's reputation for lawlessness. As a result, police officers who patrolled the MWS often tended to mistake any young woman on the street as a potential criminal, adding to the existing tensions between residents and the police force.

By creating zones of exclusion, the police force accomplished several objectives. The zone around the financial district, and the use of special squads for "clearing" criminal elements, pleased the merchant and finance class, a key constituency, particularly for Tammany-backed mayoral administrations trying to prove their reform credentials.[125] The construction of these zones also isolated the criminal element into certain defined, controllable areas, making it easier for police officers to perform crackdowns when necessary, and to easily locate their sources of extra income. But these zones of containment also, when combined with standard policing methods and common legal understandings of the period, created frozen areas within which the rights of citizens could be violated with impunity. The language used to describe the frozen zone of the Middle West Side by the police, reformers, civil authorities, and the press served to solidify the physical and cognitive boundaries of the area, and lock in, for both residents and outsiders, the common understanding of the region as "violent," "lawless," and "vice-ridden." The public perception of the Middle West Side was of an area of criminality whose residents neither deserved nor commanded the full rights of citizenship, a geographic zone of potential impunity. Similar to war zones, these zones of impunity, unbound by the rule of law, became spaces where law is arbitrary, and its application viewed by those within the zone as a violation of their positions as legal citizens and subjects of the state.

POLICING THE BOUNDARIES

IN THEIR POPULAR HISTORY of the NYPD, James Lardner and Thomas Reppetto suggest that, at its founding in 1845, the professional force was "the child of the press," created in response to sensational stories of violence and the cry to "reinvigorate the administration of criminal justice" emanating from the *Herald, Tribune,* and other papers of the day.[126] Indeed, stories of violent crime perpetrated by ethnic or racialized individuals or groups were a standard trope of nineteenth-century newspaper journalism, and never more so than during the 1890s and the Lexow Committee period in New York. Aside from stories of grisly murders and everyday violence in ethnic neighborhoods, newspapers also focused on either the ineptitude of police officers, often attributing this to ethnicity and class, or presented officers as heroes protecting a vulnerable public from lower-class ethnic miscreants. Crime reporting, the coverage of police scandals, and selective transcribing of court testimony were standard for New York dailies, helping to boost sales during a period of intense competition among papers. Reports of crime, both violent and petty, and tales of inept police work, were often used as comic filler, two- and three-paragraph reports lacking full detail, and filled with what could be termed stock characters.[127] The daily papers of the day were also major players in the partisan political battles over control of the city, with each tabloid choosing sides between Tammany Democrats, Swallowtails, upstate Republicans, and reformers, each trying to appeal to both identified audiences and the general public.[128]

The public perception of the Middle West Side as an area of crime and lawlessness, encouraged by the creation of frozen and vice zones, was solidified by the press reporting. As the completion of the Ninth Avenue elevated and the influx of new immigrants and native-born working class increased congested conditions in the area, the view of the Middle West Side as Hell's Kitchen, a place without the glamour of the Tenderloin, or even the ethnic charm of the Lower East Side, became entrenched in the attitudes of city officials and in the public perception formed by the tabloid press. Newspaper reports and journalists' accounts of the area consistently used "notorious," "crime-ridden," "dangerous," and "chaotic" in their opening lines to news stories, and residents were often described as "denizens," a word with dark overtones. Press accounts also reported—or promoted—the practice of "slumming," in which "respectable" citizens, often with paid escorts, would visit the area to experience firsthand the authentic feel of a vice zone.[129] The area, along with the Lower East Side, became a signifier of the urban problem in the United States, the "shame of the cities," as Lincoln Steffens would call it, during the period of peak immigration, standing in as both scapegoat and object of moral concern. For Middle West Side residents, occupying this position of a "dangerous" community meant that their standing as full members of the larger community of citizens was constantly in question, particularly during this period of large-scale immigration, rapid social and economic change, and shifting boundaries of proper citizenship based on race, ethnicity, and class. The story of how the Middle West Side residents, in the context of racial and ethnic citizenship, "became white" is a story embedded in the shifting boundaries of citizenship and in the changing nature of their relationship to spatialized forms of public authority.

The physical realities and cognitive understanding of the Middle West Side's various boundaries or edges, for both residents and authorities, created a zone in which relations between city officials and those they allegedly served were severely strained, with tense situations often leading to violence. While the press most often focused coverage on the violent activities of residents, the actions of police officers and their arbitrary treatment of locals contributed greatly to the tensions in the area, and helped to sustain the reputation of Hell's Kitchen as a zone of police impunity. This level of police violence, contained within the frozen zone of geographic criminality, created among Middle West Side residents an uncertainty as to their status in the wider community. When combined with a lack of services, substandard housing, and other spatial

A dead horse on the street, around 1900. Lack of services to clear streets would contribute to an area's reputation. (http://ephemeralnewyork)

conditions, as well as contemporary ideas regarding the essential nature of race and ethnicity expressed in social science and popular writings, geographically determined police violence and impunity created and sustained the conditions that encouraged incidents of violence based on racial, ethnic, and class identities. Thus understanding the geographic nature of policing and the production of this type of relational space is vital to any examination of the formation and re-formation of such place-based identities.

But, as police historian Marilynn Johnson states, "Determining the actual level of police violence at any given time is virtually impossible." New York City kept no records of civilian complaints against police until the 1950s, and even these records, given the nature of the civilian-police relationship in "high crime" areas, are often unreliable as indicators of the level of police impunity. As well, policing practices have changed over time. Arbitrary use of violence by police, as well as open graft and corruption, acceptable in 1890, had become by 1910 grounds for dismissal from the force in certain cases. Thus, as Johnson suggests, any study of police violence against citizens in a given area "must deal with the public perceptions of the problem." The perceptions of the Middle West Side residents include their own self-perception as members of the wider community of rights-bearing American citizens, a perception based in part on how they were treated on a daily basis by civil authorities and perceived by the wider public. For many in the area, the identification of their community as a zone of criminality, and as an area where police impunity was largely unchecked, meant that theit status as citizens was uncertain and shifting. The result of this uncertain and shifting status was reflected in the many ways Middle West Side residents protected themselves and their neighbors from arbitrary actions. These strategies often included solidarity along racial and ethnic lines, and also solidarity along and within perceived spatial edges.

Despite of the absence of statistical compilations of police violence, measuring public perception regarding attitudes toward authority and citizenship in the Middle West Side is possible by examining the intersection of certain statistical evidence and narrative to form a historical snapshot of the citizen–police relationship. For example, Johnson's study of brutality charges against police officers, brought between 1865 and 1894, shows a spatial pattern of violence. Over 20 percent of all brutality cases against officers were brought by alleged victims from the area comprising the Middle West Side and the Tenderloin. Of course, these numbers are problematic given that residents in poor or underserved areas with a reputation for police impunity may have been less likely than other city residents to bring charges against officers. But a more interesting aspect is the charge of "disorderly conduct" as an indicator of potentially arbitrary action on the part of officers. Johnson's study shows that the criminal charge against civilians in 50 percent of cases of alleged brutality was disorderly conduct.

The charge of disorderly conduct can cover a wide range of activity and behavior, and is a favored tactic of police officers in maintaining public order. For New York City, the Penal Code definition of disorderly conduct included such arbitrary clauses as "refuse to comply with a lawful order of the police to disperse," creates a hazard by "any act which serves no legitimate purpose," "obstructs traffic," and engaging in "threatening behavior." George Chauncey has shown in his work on the construction of gay culture in New York City that disorderly conduct charges were frequently used by police to disperse congregating groups of "fairies" and perverts.[130] The vagueness of this charge allowed officers to apply disorderly conduct statutes in almost any situation, and to act in ways that the public could perceive as arbitrary or punitive to a particular population. It is not surprising, then, that the use of disorderly conduct charges would result in acts of police violence, as citizens would be less inclined to accept the officer's interpretation of their activity as disorderly when the definition of such acts was subjective.

A measure of the potentially arbitrary nature of the application of disorderly conduct statutes can be found in the rate of arraignment and release on such charges. Taking the year 1908 as our sample, records of the District Courts of Manhattan show that in only two districts, the Second and Seventh—the Lower East Side and Middle West Side, respectively—did the number of "discharged" cases of disorderly conduct exceed those "held" or brought to arraignment. Of the 5,255 citizens arrested on disorderly conduct charges, some 3,698, nearly 70 percent, were dismissed without formal charge. This meant that, in 1908, almost seven out of ten people arrested in the Seventh District court area, comprising most of the Middle West Side, were arrested for reasons that could be viewed as arbitrary, where the arresting officer knew the charge would not stand up, or had made the arrest possibly in violation of the basic rights of the individual. In contrast, for those charges in which the definition of the actual criminal activity is more clearly defined, such as homicide, robbery, burglary, and even violations of sanitary law, the rate of persons arraigned far exceeded that of those released. The statistical evidence for 1908 clearly shows that for crimes in which the subjective judgment of the arresting officer was paramount in the arrest decision, such as disorderly conduct, suspicious persons, misdemeanor otherwise not classified, and intoxication, arraignment rates were either around or well below 50 percent. For crimes that required physical evidence and were less subjective, such as assaults, property crimes, and murder, the opposite holds (see Table 1).[131]

Table 1

Offense	Held for Arraignment	Discharged
Disorderly conduct	1,559	3,698
Suspicious Persons	0	133
Intoxication	886	916
Homicide	33	0
Robbery	194	4
Burglary	143	14
Violation of Sanitary Laws	333	40

The arraignment and dismissal rates show a telling disparity between criminal charges based upon the more subjective judgments of arresting officers and those based upon defined crimes with more objective evidentiary standards. This is not to say that police officers do not use subjective judgment when making arrests for property crimes, but only that the charge of disorderly conduct and other vague statutes are more subjective and liable to dispute, as well as being much more numerous. It suffices to point out that for crimes such as homicide, robbery, and burglary, there must be physical evidence, something stolen, misappropriated, or someone killed. The same does not hold for arrests for being disorderly, suspicious, or, in 1908, intoxicated. As Chauncey points out, being disorderly could cover a wide range of "non-normative behaviors in public spaces" including gambling, drinking, or simply congregating with no "legitimate" purpose.[132] The order to "break it up" or "move along" was often meant to either prevent illegal or illegitimate activity from taking place, or to ensure that such activity was occluded from the public eye. Groups of citizens, most often men, congregating without legitimate purpose threatened the ability of police officers to maintain their control over public space. Congregating citizens in zones of criminality could prevent officers from seeing illicit activity, and large groups of people directly challenged the violent force available to the lone foot patrol officer. The threat of arrest for being disorderly was thus a vital tool in the maintenance of public space.

Residents of the Middle West Side often contested this process of public authority. Contestation over control of public space took place in a variety of forms, from challenging basic definitions of property laws to hindering officers in their daily duties, to open confrontations with police. Though not always hostile to public authority, Middle West Side residents often acted within the constraints of their spatial environment, performing their roles as the "dangerous" classes while testing the boundaries and borders of normative structures. Their active or sometimes passive contestation over public space must be viewed within the context of their own attitudes toward public authority and the changing relationship between police and citizen. Between 1894 and 1914, the NYPD changed dramatically in its hiring, training, and daily procedures. During this period, civil service procedures were instituted in hiring in 1884,[133] and codified training processes, handgun instruction and qualification, and the Bertillon system of criminal classification were adopted during the 1890s. These changes were concurrent with changes in the demographic composition of the City and Middle West Side, and with other changes brought on by economic and political conditions.

In her West Side study *Boyhood and Lawlessness*, Pauline Goldmark writes that though most city residents displayed an "indifference" or "passive antipathy" toward the police, they "openly conspire and are actively hostile" toward them.[134] A 1906 report from the Seventh District Magistrate Court sheds light on the reasons for this perceived hostility as Magistrate Court Judge Whitman questioned the arrest of ten women in the area bordering the Tenderloin and Middle West Side neighborhoods. Of the ten, only one, an African American, was actually sentenced to the workhouse on a charge of prostitution, while the rest were found innocent for lack of evidence. The arresting officer testified that his only evidence against the other nine was that they had been talking to men. They had not been intoxicated, disorderly, nor was their any suggestion of solicitation. In Whitman's opinion, the women had been arrested so that they would be forced to seek five-dollar bonds for their release from local bondsmen, who then kicked back part of their fees to the arresting officer. Most likely, this was not an isolated incident, as the officer testified that he was acting under the orders and with the approval of his precinct captain. It is hard to imagine ten such arrests occurring in an area frequented by middle-class or wealthy residents. The fact was that the freezing of certain areas and the reputation they developed directly contributed to incidents of police impunity and to the negative attitudes of citizens toward police officers and other public institutions. But equally important, citizens treated with this level of impunity can never be certain of their status in public space, nor of their position within a larger society that was alleged to be based in a set of basic, recognized rights.

Nine women being arrested without evidence was probably not news to Middle West Side residents. What was at least new, or emerging, in 1906, was that a local magistrate would question an arresting officer's actions, and that the story would be covered in the newspaper without bias against the women. By 1906, attempts had been made, starting with the Lexow Committee, to reform the police department and improve its standards in hiring, training, and procedure. Efforts were also made to break the close ties between the police department and the local Democratic Party structure. Reform officials, such as future president Teddy Roosevelt, instituted new rules regarding the use of the club, standardization of uniforms, issuance of standard revolvers and training in their use, and civil service procedures in hiring. As well, reform organizations like the New York Bureau of Municipal Research assisted the department in instituting modern accounting practices and helped to Taylorize daily routines for street and desk officers. Standard procedures of the 1880s, such as the indiscriminate use of the nightstick in lieu of arrest, had been eliminated. Investigations into graft, political activity, and outright corruption had led to more professional standards in training and daily activity. Police reform resulted in the actual dismissal of officers, such as Thomas McCormick, known on the West Side as "Terror Mac." Famed for his use of rubber-soled boots for planting a strategic kick, McCormick was dismissed from the force in 1902 for taking kickbacks from bail bondsmen, ending a career that included twenty-seven disciplinary charges. But in spite of these improvements and reforms, police officers still found themselves in the position of being arbiters of immediate state power, and in a position open to the temptations of graft and impunity.

Thus the relationship between residents of the Middle West Side and the officers on patrol remained troubled, at best, and were often hostile and antagonistic. Though the reports of the Lexow Commission turned a spotlight on the indiscriminate use of violence by police against citizens, the main effects were a reduction of arbitrary action in more "respectable neighborhoods." New rules and procedures reduced and restricted such practices as the use of the nightstick as deterrent, but Middle West Side residents remained susceptible to arbitrary arrest, shakedowns, geographic suspicion of criminality, and outright violence based largely on their position as residents of a zone of criminality. The results of this often antagonistic relationship were twofold: on the one hand, the attitude of the police to MWS residents meant that those residents were never sure or certain of their own safety. On the other hand, the lack of respect that developed among residents meant that the police, and by extension city government, were rarely ever viewed as solutions to the areas long- and short-term problems, but were viewed as yet another difficulty to be overcome or avoided. For all the talk and efforts aimed at reform and efficiency, the geographic marking of the Middle West Side as a zone of impunity made improvements in community–police relations difficult if not impossible. As Harvard political scientist George McCafferty pointed out in 1905, "The efficiency of a police department depends largely upon the respect with which it is regarded by the populace, and this respect rests in a very large degree upon the ability and tact with which small happenings are handled." McCafferty's statement is particularly prescient for the Middle West Side, where the lack of tact and objectivity often turned "small happenings" into major events.

Small events involving the police could include anything from congregating on the street, domestic disputes, truancy, petty theft, and vagrancy to simple disputes over money owed or rents unpaid. The testimony of residents throughout the period indicates that the arbitrary handling of small happenings caused by the reputation and material conditions of the area greatly contributed to the antagonistic nature of community–police relations. West Side residents were keenly aware of the arbitrary nature of police activity, and of its ramifications. "The cops are always arresting us and letting us go again," a youth complained. "These cops will give you a bad reputation if you've never had one before in your life."[135] Complaints about arbitrary action often revolved around activity that residents considered normal, such as the appropriating of wood, ice, and other scrap materials, or the inevitable disputes that occurred between neighbors and among family members in the crowded conditions of the tenement. For the year 1909, the Bureau of Social Research listed among the offenses for which young boys had been arrested such daily activities as ball-playing, pitching pennies, shouting and singing, upsetting ashcans, and the almost ominously arbitrary "general incorrigibility." Being "hauled off" for ball-playing, loud arguments, or simple "loitering" was viewed by many residents as an excessive overreaction by police.

As a result of these attitudes, residents' resentment escalated. One police officer, in 1902, recounted the story of how his arrest of two men on the Hell's Kitchen waterfront for allegedly consuming stolen beer was thwarted by an angry mob as he attempted to walk his charges to the precinct house.[136] Numerous officers reported having things thrown on them from rooftops, people distracting them to allow a suspect to escape, and, most commonly, being met with silence when looking for witnesses to crimes. Goldmark reports that area residents "openly conspire" against police, treating them as a "natural enemy." As "hatred of the police had become a tradition," the beat officer's uncertain status, his position as the point of contact between citizen and governmental authority, was further complicated. Police were "unable to cope" with either the situation on the streets, where their efforts were consistently thwarted, but also in the courtroom, where working-class witnesses often engaged in what McCafferty refers to as "soak the cop," a deliberate conspiracy to contradict police testimony and keep a fellow resident from a fine or jail. The hardened attitudes of residents toward the police led to a condition more serious than mere hostility. As Goldmark reports, even young boys "laughed at their authority," a level of contempt running deeper than simple opposition.[137]

Adding to the antagonism was the role of police in labor disputes. In the longshoremen's strike of 1907, a long and violent labor dispute between dockworkers and several stevedore companies, the NYPD, under orders from the mayor, provided protection and escorts for strikebreakers. Police violently dispersed picket lines that were attempting to disrupt the loading and unloading of ships and repeatedly utilized loitering and disorderly conduct statutes to harass striking workers. They categorized as "riots" several attempts to block non-union workers from entering the work area, using clubs indiscriminately and causing numerous injuries. The 1907 action was just one in a long line of labor disputes on the West Side waterfront, disputes in which dockworkers faced two opponents: management and the police.[138] The role of the police in labor disputes deepened feelings of antagonism between officers and male workers, who often complained of police interference in domestic disputes, and police overreaction to incidents of public intoxication.

The consequences of the strained relationship between the police, city authorities, and Hell's Kitchen residents were detrimental to all. Police and city authorities could not count upon the cooperation of residents in pursuing real criminal activity or in attempting needed relational improvements. On the residents' part, returning to Professor McCafferty, they did not see the police and other authorities as neutral adjudicators and administrators dedicated to preserving public order. For residents, police officers, elected officials, and administrators occupied ambiguous and often confusing positions within their daily rounds. Even though many police officers came "from the neighborhood," and were of similar ethnic backgrounds[139] they often lived in slightly more upscale areas, such as Eighth Avenue above 40th Street, as their salaries and perks provided them with a higher standard of living than the typical MWS resident.[140] Some officers were friendly with those they encountered on their rounds, whereas others maintained either a cold neutrality or hostile indifference. Police were sometimes seen as allies in the inter-ethnic struggles occurring on the MWS, but even ethnic loyalty could not be counted upon. As direct enforcers of state power, and as individuals who often supplemented their official income with "extras," officers were in an ambivalent position, setting boundaries and borders of criminality and respectability as they often crossed the line themselves.

The ambiguous position of police officers and their role in producing spaces of difference is exemplified in the race riots of the summer of 1900. The incident that set off the violence was based on geographic policing, and the subsequent battles in the streets showed the kinds of anxiety and identity confusion that police impunity produced. Many historians and urban theorists have pointed to the ways in which identity groups redeploy space to produce areas where they can perform solidified signifiers of their particular subgroup, be it ethnic, racial, gendered, or sexual, but here authorities and their deployments of power threw those very identity performances into question.

On August 13, 1900, at 2 a.m., as May Enoch, an African American, waited outside McBride's Restaurant at the corner of 41st and Eighth, the Tenderloin–Hell's Kitchen boundary, for her common-law husband, Arthur Harris, plainclothes officer Robert Thorpe made a decision based on both geography and racial stereotyping. Assuming that Enoch must be a prostitute, or seeing an easy opportunity to grab a cut from bail money, Thorpe attempted to arrest Enoch on charges of solicitation. Emerging from the club and seeing his wife being treated roughly, Harris struggled with Officer Thorpe, who delivered several blows from his club before Harris "cut him twice."[141] Thorpe died the next day in Roosevelt Hospital as Harris fled the city to his mother's home in Washington, D.C. While funeral preparations were made, and the search for Harris commenced, racial tensions in the Hell's Kitchen area increased, urged on by groups of mainly young men of Irish descent. The tensions were heightened by the increased presence of African American renters within the official Hell's Kitchen area and its surrounding neighborhoods to the north and east. These tensions were exacerbated by the preference of many landlords for African American tenants, who were considered more reliable, and whose limited housing options allowed landlords to charge them higher rates. As a newspaper article of the period states, black tenants were renters "who could be depended upon to pay their rent, who do not damage property, and who are orderly and peaceable."[142] It also did not help that Officer Thorpe was scheduled to marry the daughter of local precinct captain Charles Cooney, and that Thorpe's family was from Hell's Kitchen.

An argument between a friend of Thorpe and an African American passing in front of his home on August 15 escalated into violence, as white mobs hunted African Americans on the streets. Many observers believed that the incident in front of Thorpe's home was purposefully set up to trigger violence, a scenario reminiscent of white violence in other U.S. cities during the period.[143] Numerous reports and complaints of police inactivity or active participation were lodged not only by African American victims of attacks, but by white witnesses appalled by the lack of protection afforded to the black residents. Despite the fact that emergency rooms in the area quickly filled with black patients, no arrests of white attackers were made, prompting a night magistrate in the 37th Precinct to demand, at midnight, that at least some attackers be arrested. For the evening, a single white male, a teenager named Frank Minouge, was arrested for interfering with police business, not for assaulting black citizens.

The violence resumed on the evening of August 16, at first following the same pattern of white mob violence and police indifference or direct participation. As the violence spread, Chief William Devery, under pressure from press coverage and from outraged citizens who witnessed the treatment of blacks by police officers, decided to clear the streets by using a police charge. The *New York Herald* reported that the police, led by Captain Clooney, moving down Eighth Avenue, "clubbed everyone in sight who refused or delayed in getting off the Avenue." As the *Herald* also reported, "By midnight, the cry was no longer 'Damn the Negroes,' but had become 'Damn the Police,' as club-swinging officers acted with complete impunity, violating the rights of both black and white residents indiscriminately."[144] *New York Herald* stories reported the "shock" white rioters felt when the police turned on them. As the white riot turned into a police riot, the racial solidarity felt by white residents toward the police force melted away, producing confusion and disillusionment among white residents.

The white riot of 1900 was not an isolated incident, nor did it follow any typical pattern of public violence in the MWS. In 1898, several days of rioting resulted from rising tensions over blacks moving from their enclaves near Seventh Avenue and 30th Street to traditionally white Irish neighborhoods to the north and west. Encouraged by landlords who claimed that "Negroes pay their rents regularly, and many of the whites do not," black renters moving into the Ninth Avenue area increasingly organized themselves in self-defense. In the 1898 riot, which lasted for four days in early August, black gangs fought white groups to a virtual standoff, while police, without the 1900 killing of one of their own to prejudice them, arrested indiscriminately, although many witnesses claimed they were biased in favor of the Irish. While the officers on the scene inflicted violence on people on both sides of the dispute, at least six white residents were arrested and charged with assault, and most others were charged with misdemeanors such as disorderly conduct. Subsequent newspaper coverage laid much of the blame for the rioting on Irish gangs, who were accused of "terrorizing the tenements."

Other outbreaks of public violence in Hell's Kitchen involved youth gangs from different streets battling each other and the police over turf, and sometimes, it seems, for entertainment, and events such as the "Austrian" riot discussed earlier. Contemporary reports on the street gang riots make no distinction among racial or ethnic groups, and surveys show that street blocks, though often predominantly Irish, were ethnically mixed. Accounts from residents indicate that street gang membership often crossed ethnic lines, and ethnic solidarity seemed a fallback position for certain types of dispute. Although women also participated in public violence, they seemed to act as protectors of the neighborhood against police officers, trying to keep their men from costly arrests. In all cases of public violence, however, both police and reporters refer to the protection of "respectable" citizens, indicating class differences among residents. And all residents regardless of social position were susceptible to arbitrary action by police officers.

The 1900 riot exemplifies the spatial production of difference generated by police impunity. As racial solidarity with white police officers turned into arbitrary police action, Hell's Kitchen residents experienced the ambiguity of their own social positions, and felt the insecurity of not knowing what rights they possessed as citizens. Under such circumstances, how was an individual, be he an Irish dockworker, an African American woman waiting for her husband, or a Croatian mason sitting in a bar, to know what rights they possessed and when or if they had crossed a boundary where their status shifted? Further, though it was possible for citizens to experience arbitrary treatment in other parts of the city, the geographic designation of the MWS as a zone of criminality made the expectation of arbitrary treatment a fact of daily life. In response to such expectations, residents of Hell's Kitchen often relied upon a shifting array of solidarities to defend themselves and make claims to their rights as citizens. These solidarities, based upon race, ethnicity, occupation, and class, though fluid in the daily routine, coalesced at periods of heightened tension, emergency, or violent confrontation. Daily routines, such as shopping, bar drinking, or rooftop kite flying, were activities that allowed for boundary crossing, though identity lines, particularly ethnic and racial, were still marked by material and discursive signs. Racial, ethnic, and class solidarities appeared during periods of spatial dispute and in the face of arbitrary treatment at the hands of authorities. But as the events of August 1900 showed, not even racial solidarity could protect local whites from arbitrary treatment and violation of their basic rights as citizens.

Spatial policing and arbitrary police conduct mirrored the experience of Hell's Kitchen residents in other encounters, both with city authorities and in the marketplace. Hell's Kitchen residents experienced a lack of public provision such as trash pickup, upkeep on the dock areas, and a lack of open space and playgrounds for children. They experienced arbitrary treatment at the hands of employers, who dismissed workers without cause in the knowledge that an excess labor market could fill the unskilled work performed by many residents. When workers organized for their own protection, as the dockworkers did, they often witnessed city authorities and the police force siding with their employers in times of dispute. In the area of housing, constant disputes with landlords seeking higher rents, and living in substandard housing with little in the way of legal rights, made housing another area of uncertainty. This level of uncertainty, though caused in part by ethnic, racial, class, and gendered signifiers, was also the result of geography, as rights and levels of citizenship were dependent upon spatial location.

The uncertainty or difference produced by the constructed geography of the Middle West Side played an important role in both resident and outsider understanding of who was and was not a "proper citizen" at the turn of the century. David Roediger and others have pointed out that racial and ethnic identity, questions of "whiteness," played a vital part in determining the legal and cultural standards of citizenship during the period of mass immigration: "The legal equation of whiteness with fitness for citizenship shaped the process by which race was made in the United States."[145] Roediger examines how "new immigrants," those from southern and eastern Europe, occupied what he terms "in-betweenness" in the early twentieth century in their struggle to gain acceptance as full "white" citizens and to gain the advantages "of the political rights that whiteness conferred."[146] Oftentimes defining oneself as properly white meant distancing one's group racially from African Americans, thus occupying a status somewhere between the old-line Anglo-Saxon whites and American blacks, a position first occupied by Germans and Irish Catholics in the mid-nineteenth century. For Roediger, this was the cultural space occupied by "greasers, dagos, and hunkies" in the immigration period, as the Irish and Germans had already achieved a certain level of legal and cultural whiteness. Questions of race, ethnicity, proper behavior, and origin were crucial to the period debates regarding which groups constituted proper Americans.

As the debate over citizenship and belonging at the national level revolved around questions of race, ethnicity, and national origin, Middle West Side residents experienced these debates as questions of space and geographic determinism, and found themselves occupying a geographic "in-betweenness" similar to what Roediger describes. In terms of legal citizenship, the majority of Middle West Side residents qualified as citizens, or if newly arrived, potential citizens.[147] Regarding whether their demands and rights were seriously considered by authorities and other residents, these issues of cultural citizenship were often questions of space and geography. In other words, especially for the urban poor, physical location of work and residence often determined if one was taken seriously. On the local level, Hell's Kitchen residents experienced themselves as American citizens and performed their roles as ethnic, racial, gendered, or class-based groups and individuals based on common conceptions of identity markers and as inhabitants of an area bounded by reputation and outside perception. As the behavior of New York City police officers illustrates, the position of Hell's Kitchen residents as proper citizens with a protected set of rights was an open question during a period when ethnic, class, and gendered groups were struggling to gain such privileges. The status of Hell's Kitchen as a frozen zone of criminality where police acted with impunity and public authorities were unresponsive meant that those who traveled the paths, districts, edges, and borders of the Middle West Side experienced diverse forms of uncertainty and relied upon a variety of shifting solidarities in response.

For a resident like May Enoch, waiting in the early morning hours on the edge of the Tenderloin for her husband, the uncertainties took multiple forms. As an African American, she would have been aware of the racial tensions in the area, but perhaps would have felt comfortable at 41st and Eighth, where blacks and whites seeking entertainment often mixed easily. Enoch was probably more aware of the fact that as a woman in such an area, she was subject to arbitrary arrest as a prostitute, based on geographic location and the attitude of local patrol officers. For a woman like Enoch, such arbitrary treatment often meant either paying bail and seeking legal help, or perhaps losing a job because bail could not be immediately arranged. Such individuals might also suffer undeserved damage to their reputations, or the arrest could cause tension within their relationship with their partner. Such uncertainties regarding a basic right to be on the street in a certain location at a certain time might cause some women to not take types of employment that might otherwise pay well, but place them in positions of precariousness. As a woman, an African American, a member of the working class, and a resident of Hell's Kitchen, May Enoch fell somewhere in between the status of a person with full rights and one with no rights at all.

Others, such as Frank Minougue, the teenager arrested on August 15, 1900, for "interfering with a police officer," also lived in a life world somewhere in between full citizen and non-citizen. Being in the wrong place at the wrong time could result in similar consequences, such as losing a much-needed job for failure to make bail. During periods of public violence, the innocent act of simple curiosity—seeking entertainment by finding out what was happening on the street—could result in arrest, and by many accounts violent treatment at the hands of police. By the same accounts, a simple domestic dispute could result in a clubbing if the wrong officer responded. Seeking entertainment in a nightclub during the periodic vice raids brought on by reformers or to impress political operatives, could result in the same. Residents often had no way of knowing if they had crossed a boundary, either physical or cognitive, when dealing with police and public authorities. Living in an area where basic rights were freely and frequently violated created the conditions of uncertainty that led local residents to question their own position as citizens within the larger culture, and often caused them to fall back upon the racial, ethnic, class, and gendered solidarities that provided both protection and mental solace.

CHAPTER FOUR

Housing and Visible Spaces

Vegetables grown to maturity were proudly brought to the Commissioner by the young gardeners of both sexes as an evidence of their success, and it is believed that the children participating have become imbued with a love of nature and an appreciation of the beauties of the park, as they could have become in no other way.
—DEPARTMENT OF PARKS, REPORT OF 1902

THE CONSTRUCTION OF DeWitt Clinton Park, in the northernmost reaches of Hell's Kitchen, was intended by its supporters to provide open, visible space in a region of cramped occlusion. The original park plan called for sloping hills, rock gardens, boys' and girls' playgrounds, and winding footpaths on its seven acres. DeWitt Clinton was the culmination of some thirty years of effort by various concerned citizens to install open park space on the far West Side of Manhattan. Disputes over location, condemnation of existing structures, and funding had long delayed construction, which only commenced with the passage of special legislation by the New York State Assembly in 1900.[148] Construction began in 1902, and would not be completed until 1905. But so impatient were local reformers for the park to open that the Children's Farm Garden, pictured above, was opened through the efforts of Mrs. Henry Parsons while the land was still being cleared. For Mrs. Parsons, as for Lawrence Veiller and other reformers, open, proper space would provide the basis of no less than "rational human existence" for the area's young people, converting the space of unregulated development into the planned places of community. According to the Parks Department report, the results were immediate and apparent. The mere connection of children with soil, growing things, fresh air, and open space caused a near-miraculous transformation.[149] Those who were once blind, could now see. What was once unregulated space was now a community place.

For many Progressives working in America's cities, the provision of proper space for the urban poor was both obsession and framework, and helps to differentiate their efforts from previous urban reform traditions. Progressive visions of urban space were also intimately bound up with the larger concerns of space and place that occupied American culture and politics during a period when fierce debates were taking place about overseas expansion and "foreign" populations.[150] For many Progressives, spatial reform represented the turn to rationality in urban design and a rejection of the anti-urbanism of the American philosophic tradition.[151] Acquiring new ideas and attitudes, from their studies abroad in Europe and their own interactions, this new breed sought to create a different sense of *place* based upon ideas of abundance, rational planning, and a faith in their own efforts. Conceiving cities not as sites of decay, they sensed the danger of impending catastrophe but also the democratic possibilities, exemplified by Frederic C. Howe's now-famous *Hope of Democracy*.[152] As historian Daniel Rodgers notes, for Progressives throughout the industrializing West, "cities stood at the vital center of the progressive imagination."[153] The imagined spatial communities of the Progressives, proper visible urban space, held out the potential for a new America, and a new American.

But if DeWitt Clinton Park represented the triumph of urban reform in 1905, it also stands as a cautionary tale. The park suffered from a lack of maintenance from its earliest days to the present. It took over thirty years to achieve financing, and by the 1930s the park had lost over one-third of its original size to other projects, such as the West Side Highway; the park's Grand Pavilion, fallen into disrepair, was used as landfill by 1939. The park was part of a major effort by reformers like Lawrence Vellier to combine park provision with new model tenements and public bathhouses, creating a spatial city of openness and visibility that would reduce crime and encourage proper citizenship by providing a sense of place and

Children play on the swings in DeWitt Clinton Park, 1911.
(http://www.nycgovparks.org)

community.[154] The grand design of the reformers would collide with the needs of the real estate market, as multiple efforts to construct model housing would either fail outright, or wind up co-opted by private interests. The imagined spatial communities overlapped the economic scales of a local and regional regime of accumulation and development. Reformers and urban planners, in their desire to create a community place out of urban space, produced instead a corpus of knowledge that would influence action concerned with the "urban problem" for several generations. At the same time, much of what we know of urban reform and the Progressive urban movement was, as I have tried to demonstrate, produced by the urban populations themselves, and the spaces they created and transformed. One of the lasting legacies of the vision of turn-of-the-century reformers was the division of urban zones into "space" and "place," which remain among the most enduring specters of their efforts.[155]

Contemporary geographers and spatial theorists typically draw the distinction between abstract space and particular place.[156] "Whereas space refers to the structural, geometrical qualities of a physical environment, place is the notion that includes the dimension of lived experience, interaction, and use of space by its inhabitants."[157] In this distinction, place represents community, the carved-out redeployment of impersonal spatial processes and the construction of demarcated zones of interaction and performativity.

As urban geographer J. Nicholas Entrikin states, "Our relations to place and culture become elements in the construction of our individual and collective identities."[158] For these geographers and theorists, place is the location of agency, the context in which human individuals and aggregates construct their identity boundaries and often advance or defend political, cultural, and social claims on the larger community. It becomes the site of personal interaction and, in functionalist logic, also becomes the template for that interaction, an essentialized and naturalized "condition of human experience."[159] In trying to synthesize the abstract generalities of space with human agency, action "takes place" in specific locations, becoming what Leopold Von Ranke describes as "things just as they really happened." Through this process, the impersonal workings of natural, cultural, or market forces as spatial processes are given specific locality, which, as Mikhail Bakhtin states, "transforms a portion of terrestrial space into a place of historical life for people, into a corner of the historical world."[160]

Although they seldom employed the discourse of space and place, many Progressives deployed conceptualizations of urban and regional space as community, the physically and psychically located context of proper interaction. The idealist philosopher Josiah Royce defended regional provincialism and localism as "loyalty to community," a countervailing force against the ravages of impersonal modernity. Urban reformer Frederick Cleveland promoted a "socialism of intelligence," where local citizens spend their free time investigating neighborhood issues such as trash disposal and petty crime, and spend vacations doing comparative studies of other locales. Frederick Howe advocated for "home rule" for cities, freeing urban government from the corrupt and heavy hand of state and federal officials. And urban Progressives such as Felix Adler and Lawrence Veiller, working in New York City, played an active role in the production of model tenements and small parks in an effort to create the "place" of community among a collective population that they observed as suffering from its lack.

This chapter sketches what "takes place" within the matrix of spatial restructuring by utilizing the concept of *heterotopia*: the imagined space of representational production. In so doing, my intent is a critique of the concept of place, as used by geographers and spatial theorists. Many spatial theorists utilize the concept of place to draw the distinction between abstract notions of space and localized uses of the built physical environment. For instance, Logan and Molotch distinctly separate their analysis of urban spatial production into the exchange value of market space and the use value of lived place.[161] Doreen Massey perceives place as the relational location of daily life, including such diverse locales as the workplace, home, and tavern.[162] Here, place is the distinctive, at times over-romanticized location of redeployment. For most of these spatialists, urban space is produced at the level of uneven geographic development, where production at scale determines the unequal allocation of productive investment. Within this space, identity groups use forms of social solidarity to "carve out" or "create" the places where the performance of identity signifiers are either protected by group solidarity or accepted as the result of the non-space nature, the utopian essence, of the redeployed space. In this schema, what takes place is the claiming of space from an abstract whole to create the place of local performativity and identity, such as queer space, workers' space, and gendered space.

Though not denying the overall utility of the concept of place for thinking through urban spatial processes, I contend that place also can sometime serve to deprive urban theory of a powerful tool for understanding the production of space. Place can, of course, provide a necessary means to understand, for example, how young women turn the space of production of the sweatshop into a place of intimate communication between workers toiling in restrictive space. Thus, what place offers is a way of understanding the deployment of abstract space into the reproduction of existing social categories, such as "single young female factory worker." However, the concept of place also assumes a "natural" condition among human aggregate populations, positing place as a desired norm whose absence produces alienation and a desire for something other. Most theories of place reproduce common understandings, where abstract, impersonal, or planned spaces are contrasted to organically created places."[163] In this dichotomous relationship, human actors only feel that they belong to the particular place of experience, and only then can appreciate their relationship to larger whole. Although I do not call for the elimination of place as a category of geographic analysis, I suggest that the production of place is better understood through the concept of heterotopia.

Michel Foucault, in his lecture "Of Other Spaces," presented heterotopia as both a "real" physical "site," and as spaces that contain or inspire "the reserve of the imagination," defining in their heterotopy all other "fixed" spaces. He views heterotopias as modern sites differing from the "emplacements" of medieval society, as modern heterotopias define not placement but "a thing's movement" and stability, as settled places, as "movement indefinitely slowed down." Heterotopias are, for Foucault, places of crisis or deviation from the norm that juxtapose "in a single real place several spaces, sites that are incompatible," rest homes in a society that lionizes youth and movement. Importantly for Foucault, they are sites "linked to slices of time" that solidify the logics of spatial conception into actual physical sites that then function economically and culturally.[164] Much is valuable in Foucault's explication of heterotopia, but I offer a perspective that views heterotopia not as a physical site, but as an imagined geography, grounded in the physical built environment, that incorporates the accumulated temporalities (or history) of those who participate in its production. Heterotopia offers a view of the spatial production of place that differs markedly from the humanist geography conception and should be understood as the non-physical space of true redeployment. Neither emancipatory nor deviant, heterotopia is the representational space that is "opened up" when actual, physical space is in its process of restructuring.

For the Middle West Side, this opening up occurs under historically contingent conditions specific to the period, since the restructuring of the physical environment of Hell's Kitchen occurred within the framework of a variety of spatiotemporal processes. The heterotopic imagination, literally the multiple sites of the virtual, is constructed not, or not only, as a top-down process of power, but as a mutual process of horizontal production, where differing boundaries overlap to produce different visions of futurity. So for those dwelling in Hell's Kitchen, place represents not the stability of belonging or the romantic notion of home, but the contradictory and often incommensurable locations of conflict.

IMAGINED SPATIAL COMMUNITIES

IN HIS MAGISTERIAL *The Iconography of Manhattan Island* (see illustration on page 75), Isaac Phelps Stokes demonstrates the progressive obsession with space. Meticulously detailed and covering nearly every instance of spatial restructuring from the colonial period to the early twentieth century, Phelps Stokes's work reflects his concern with urban housing and spatial reform. Like so many others of his period, Phelps Stokes combined an interest in history, architecture, and political economy with active work such as the creation and maintenance of model tenements for New York's working poor. His minutely detailed map of Manhattan in 1908 is both an impressive accomplishment in research and a textual argument for visibility and open space.

Phelps Stokes's obsession with spatial change and restructuring is seen not only in his map-making but in his actions as a proponent of such Progressive ideas as model tenement housing. Mapping the city, charting its spatiality and identifying areas of congestion and occlusion, led Phelps Stokes and others to a critique of urban planning, as well as to envisioning the urban possible.[165]

Urban reformers at the end of the nineteenth century developed conceptual understandings of urban space and spatial processes that distinguished them from previous generations of social thinkers. Although a diverse group, reformers during the period of immigration between 1880 and 1924 came to view the city as the nexus of change, leading to a new America, or as Simon Patten termed it, a "new civilization."[166] Though not every person who can be classified as a reformer was obsessed with space, most of those concerned with urban development had an obvious interest in spatial processes.[167] Differences in approach, attitude, and background conditioned urban reformers to view spatial processes and urbanization in a variety of conceptual frameworks. For some, like Lincoln Steffens, the congested conditions and lack of government services made cities the "shame" of America, the potential grounds for disaster. For others, like Frederick Howe, the city was the site of hope, the potential turning of space into place, where newly arrived immigrants would assimilate into the American way if provided with proper guidance.

Whatever side one fell on in the debate over the future of the city, nearly all urban progressives saw and acknowledged the inevitability of the city in the industrial system, and recognized at least the potential of the city as the site of synthesis. By the early 1900s, many Progressive planners were determined to alter the view of the city from a discrete location to an urban node within a larger "metropolitan" region of economic, social, and cultural production. Further, as the very word *progressive* implies, they looked to the future in both theory and practice. It was a future in which the processes of industrialization and urbanization were a given, and where any "progress" by necessity included transformed urban space as a key component of the imagined. When urban Progressives imagined the future, the activity that shaped their actions, they conceived of urban spaces transformed into historical place, locations where previously marginalized inhabitants could enter the march of progress within the context of their own localized communities. For reformers like Phelps Stokes, Edward Veiller, and others, the transformation of space into place, of the unregulated areas of crowded tenements into proper communities, would come about through planned, regulated spatial restructuring.

Historian Benedict Anderson, in *Imagined Communities*, describes the process by which individual members of the nation conceive of a coherent national body. Anderson states that though members of the national body may never know each other, "in the minds of each lives the image of their communion," brought about through an imaginary shaped by various media of transmutation.[168] Many reformers shared a vocabulary developed at European universities and in the growing centers of U.S.-based social science. They also communicated with each other through a discourse of reform that included statistics, charts, images, and theoretical concepts. Key to this procedure was forming the image of what a properly planned urban environment would look like—imagining the community in order to bring it into being. Like Anderson's national peoples, "communities are to be distinguished . . . by the style in which they are imagined." Arjun Appaduri has pointed out that the "historically situated imaginations" of such groups as city planners and reformers form "imagined worlds" and create a variety of "scapes." Along with Appaduri's contemporary identification of "mediascapes" and "ethnoscapes,"[169] I add "reformscapes" to describe the progressive imagining constructed in journals, fieldwork, reports, and schools that spurred restructuring.

Housing and Visible Spaces

What takes place in zones of overcrowding and occlusion is the attempt to restructure the unregulated space of industry, accumulation, and rent profit into the place of community. In so doing, Progressives and urban planners hoped, in the words of Frederick Howe, to "treat the city as a unit, an organic whole," in order to "secure the orderly, harmonious and symmetrical development of the *community*."[170] Reform advocate Benjamin Marsh refers to the "right of the citizen" to live in safe, secure communities in defending the move for public action in restructuring urban space.[171] Whether they were, like Howe, Progressives obsessed with issues of spatial provision and taxation, or, like New York's Calvin Tompkins, planners concerned with more material engineering issues such as transport routes and sewage, they shared a general concern regarding city planning, and a common assumption about the importance of proper intervention in the preservation of democracy.

Recent scholarship of Progressive urban planning has suggested a variety of ways of classifying the "craze" for intervention that marks the period between 1880 and 1920, and hits its peak around 1910. Historian John Hepp, in his work on Philadelphia, rejects Robert Wiebe's "search for order" thesis, and posits planning and reform as part of a middle-class project to transform everyday life through "a continued faith in progress" and "a scientific worldview." Robin Bachin's work on Chicago's urban history connects planning and reform to a specific, historically located connection between "urban space and civic culture" that sought to "transcend the boundaries of ethnicity, race relations, and class" through a progressive vision of "order, respectability, and civic identity."[172] Along similar lines, David Ward and Olivier Zunz, in their collected volume *Landscapes of Modernity,* view urban restructuring in New York as the struggle between forms of secular rationalism and cultural forms of pluralist diversity. Though all three works are correct to focus upon what the intended effects of urban planning were, they miss several key points regarding reform, planning, and the forces of a scalar modernity at work in the urban environment.

In *The Urban Revolution,* French theorist Henri Lefebvre suggests that to understand the transformations of the late nineteenth and early twentieth centuries, we must trace the change from industrialism to *urbanism,* in terms of both the physicality of the city as a nodal point of production and urbanism in terms of what Lewis

Mumford suggested as "an archive of knowledge." For Lefebvre, urbanism refers not to a "way of life" or mode of existence for the city dweller, but rather to the production of knowledge systems that frame the increasingly global nature of capitalist production and encourage or inhibit different ways of living in and moving through the world. This transformation is a historical process, a shift in modes of thought and materiality, brought about by the settlement of town–country disputes by a fully developed industrialism. The efforts of progressives and planners to reshape the physical environment of the city, and thus the civic environment of the citizen, can be viewed, I suggest, within this lens of urban space and urbanism as the production of a mode of knowledge about economic and cultural processes and the collectives of human beings that drive those processes along. Architecture and urban form become, in this sense, what Manuel Castells calls the "organizational logics" of a cultural system of knowledge production, what architect Keller Easterling terms, through Felix Guattari, "the 'techno-scientific semiotics' that are stored in operational strata of organization and practice."[173] These organizational logics, embedded in the planning process and actual structures, emit a certain knowledge regarding urban processes, which, among other things, gathers to its own logic the ability to classify urban collective groups by naming them as, first, the "poor" and "dependent," then as the "dangerous" and as the "working poor," "slum dwellers," and various classifications that frame and define the "urban problem."

The process through which this production takes place is both scalar and spatial, occurring at the level of actual, physical environment, yet starting with the conceptual. The construction of urbanism as an epistemological system, as a mode of knowing, begins with the melding of past, present, and future that guides the conceptual vision of an altered space that will with hope "produce" rational citizens, correcting the lack of imagination of the past while securing the urban, and the Republic's, future. As with all conceptualizations, the landscape of reform begins with the imaginary, the reform vision of a better present and future.

Ebenezer Howard, one of the founding figures of the urban planning movement, states in his work *To-Morrow: A Peaceful Path to Real Reform,* that the reader is "asked to imagine an estate" embracing a certain acreage:

Housing and Visible Spaces   131

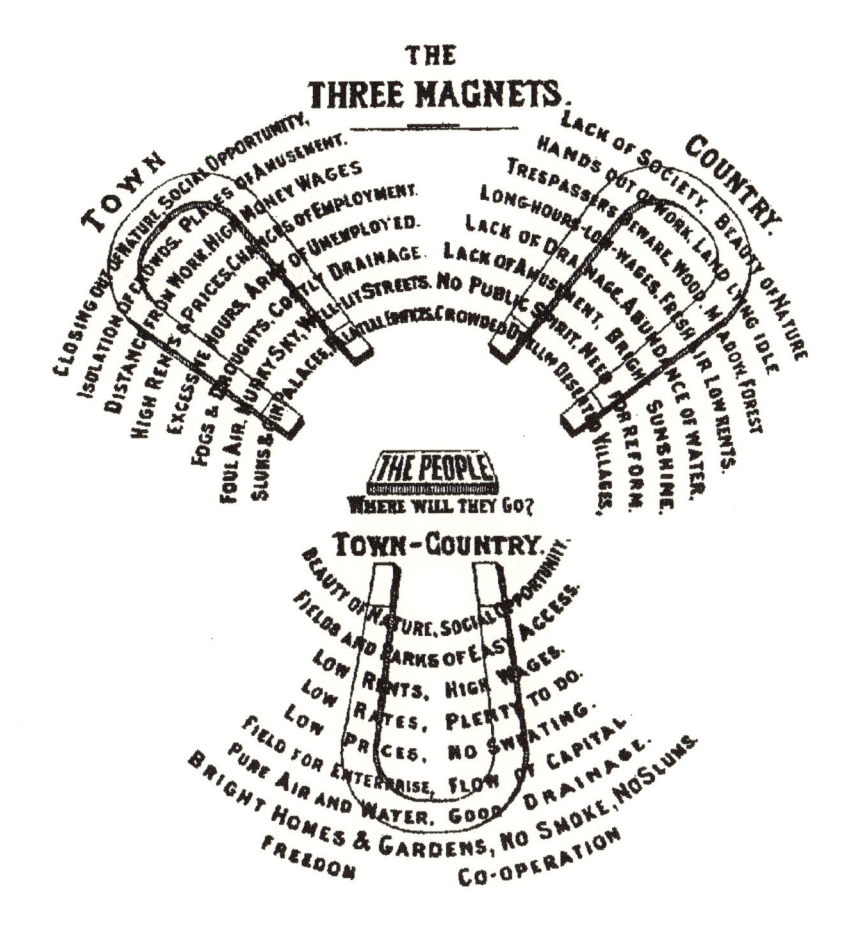

Ebenezer Howard's spatial imaginary. (From *Garden Cities of To-Morrow*)

For Howard and others involved in the planning field the vision that constructs the space of organizational logics begins with the imagination, the mental picture of the reformscape.[174] Howard, no real supporter of the actually existing urban, imagines first an empty space, a non-space utopia to be filled by the planning vision. More pragmatic planners, those invested in actual cities, still started with Howard's basic building block, the imaginary vision of the possible. Both visions were the antithesis of the actually existing urban, as each relied on what the real city could not be: light, expansive, open, circulating. Whereas Howard was mainly concerned with building his garden cities from carefully selected empty spaces, urban reformers imagined their spaces of light, expansiveness, and openness within the cramped, occluded cities of industry and commerce. For reformers and planners such as Ernest Flagg, Frederick Howe, Calvin Tompkins, Benjamin Marsh, and others working in New York and similar urban regions who were inspired by Howard's garden cities, and the "White City" of Daniel Burnham, the trick was to reconcile their conceptualization of the new, open, light city with the existing conditions of occlusion.

In his critique of the New York City grid plan of 1811, architect and planner Ernest Flagg locates the problem in the cramped urban imaginary of the city commission. The problem starts, states Flagg, with the "narrow workings of the minds" of the 1811 commission, producing "hopeless monotony" through their strictly "utilitarian" vision.[175] Like other planners, he draws a direct connection between built space and "character," explaining how the monotony and utility of the 1811 plan shaped the "habits and customs" of the urban population. As the design architect of the Singer and Scribner skyscrapers in Manhattan, and a student of the French Beaux-Arts school, Flagg was no stranger to grand designs and grandiose imaginings within the reformscape. He was a savage critic of tenement design and the lack of tenement regulation, and participated as judge and designer in multiple tenement competitions. Though many urban planners were willing to submit to the restrictions governed by preexisting conditions, Flagg allowed his reformscape imaginary full rein, envisioning completely restructured cities, including his daring suggestion in 1904 of creating a European-like thoroughfare running the length of Manhattan, effectively dismantling Central Park.

Flagg's bold vision of a redesigned city, meant to "break the bonds" imposed by improper planning, illustrates some of the tensions involved in the reformscape vision. For Flagg, the city of the future was to be thoroughly modern in design and function, combining the latest methods of architecture and infrastructure with a progressive politics that integrated the city into a functioning whole. In his focus on both the macro and micro spatial scales of the city, Flagg joined the growing group of planners for whom no aspect of the urban stood in isolation. *All was movement.* In his designs of model tenements, people, air, light, and sewage moved in harmonic balance to produce a fully functional and aesthetically

Housing and Visible Spaces 133

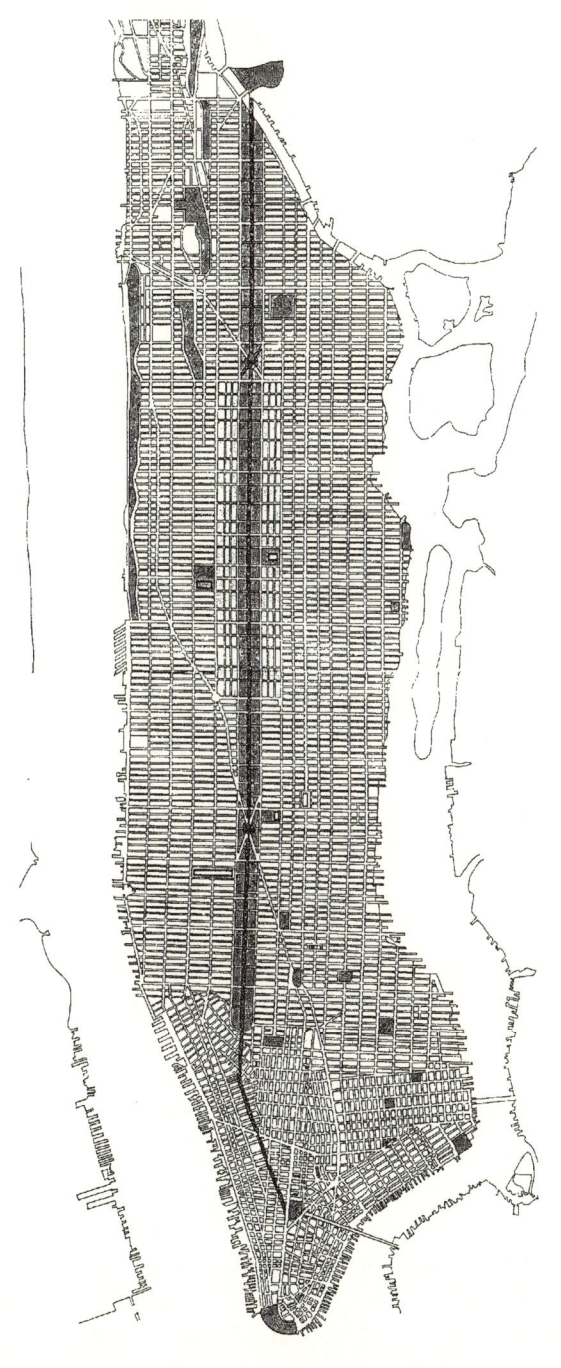

Ernest Flagg's audacious plan for Manhattan, which included the elimination of Central Park! (From "The Plan of New York and How to Improve It," *Scribner's Magazine,* August 1904)

pleasing totality. In a similar vein, his design for the removal of Central Park, and the reconstruction of Manhattan's major arteries, promoted constant, unobstructed motion. His critique of the current state of Central Park included the idea that the park no longer served its original function as ornament and pleasure ground, but now served as a major impediment, a "barrier" obstructing flows of people and traffic. Its original design, according to Flagg, was too "naturalistic," not formal enough to coexist in "harmony" with the surrounding buildings and streets.[176] Along with many of his contemporaries, Flagg was deeply influenced by planning and design ideas from Europe. His conception for a new, open Manhattan called for "wooded avenues that one finds in the great cities of Europe" with "long stretches of grass, avenues of trees, and gardens" that would be accessible to "all the people."

The dominant trend in U.S. city planning at the turn of the century countered the great traditions of American anti-urbanism and exceptionalism in its nearly slavish desire to re-create cities based on the radial boulevards of European capitals. Whether they had studied in European universities or simply visited European cities and attended international conferences, progressive planners became enamored of the European model of planning and the tradition of state and local control over urban development. As Daniel Rodgers demonstrates, the industrial cities of Europe and the Americas, and their problems, "formed a world of common referents,"[177] a language that often traversed the distance between Europe and the United States. Planners became fond of employing both empirical comparisons and rhetorical flourishes when calling for more direct control of the growth of urban areas in the United States. In the former tactic, architect Frank Koester, writing in 1912, favorably compares the population growth of German cities to U.S. cities, equating this growth with progress based on the popular German model of careful planning. He cites the "superior appearance" and "harmony" of cities such as Breslau, Dresden, and Chemnitz, admiring in particular the ability of German-planned cities to open their spaces to promote movement. Koester cites the "arrangement of traffic canalization, location of factories, the easy movement of products" as just some of the benefits of planning the Teutonic way.[178] Comparison with European cities extended to all aspects of planning, including Flavel Shurtleff's lament on the U.S. legal system, whose protection of the rights of private property had long been a bedrock of American exceptionalism. Shurtleff, a Boston attorney who specialized in urban legal issues, lauded the English Town Planning Act of 1909 as an exemplary tool for "conferring exclusive authority" on the local "government board" to have "wide power and sole authority" over housing development, something Shurtleff deems unlikely to occur in the United States due to legal tradition and precedent.[179] Shurtleff's comparison echoed Herbert Croly's lament of 1907 that the "relatively modest" recommendations of New York's City Improvement Commission had been, upon its reception in 1904, "for all practical purposes, a dead thing" due to its demand for public control of development.[180]

Empirical comparisons that favored the European model were joined with rhetorical flourishes that praised the aesthetics of the great cities of the Continent while disparaging the failure of U.S. planning, particularly its lack of "vision" and conceptual imagination. Frederick Howe, the advocate of the city as the "hope of democracy" and the urban as a necessary and vital component of the modern, laments that U.S. planning has not become an "art" as it has in Germany, mainly because it "does not visualize the city as a unit in all its relations," as a nodal point in an increasingly complex system of production, distribution, and consumption. For critics like Howe, Benjamin Marsh, and Charles Lamb, the failure of imagination and lack of creative vision included the inability of real estate developers, politicians, and engineers to visualize the scalar interactions of a global city. As early as 1876, Frederick Law Olmsted envisioned New York's future within the context of a global city, not built for the temporary exigencies of accumulation and convenience. Olmsted writes: "So far as the plan of New York remains to be formed, it would be inexcusable that it should not be the plan of a Metropolis; adapted to serve and serve well every legitimate interest of the wide world; not of ordinary commerce only, but of humanity, religion, art, science and scholarship." This will only come about if the city is planned with "the directed attention to the purpose" as evidenced in "London, Paris, Vienna, Florence and Rome."[181] Twenty years later, his son, John C. Olmsted, would liken the city engineer to the "family physician" whose responsibilities include maintenance of the city's arteries to ensure the flow of traffic just as the doctor ensures the flow of blood.

The European influence on the reformscape vision extended to the connection of spatial provision and democracy and citizenship, concepts crucial to the melding of planning with Progressivism. Benjamin Marsh, secretary of New York's short-lived Committee on Congestion and Population, used his study of European planning models and his attraction to Fabian socialism to advocate for the "right of the workingman to leisure" and the right to 'live under conditions that do not impair his health or efficiency." A single-tax adherent in the Henry George mode, Marsh was a tireless advocate for the working populations of U.S. cities, and drew clear links between proper planning and the "awakening of a sense of civic responsibility" in the urban populace. His "five distinct functions" of town planning are worth quoting in their entirety, both for what they say regarding Progressive planning priorities and the reformscape vision, and for how they anticipate the zoning laws that would soon govern U.S. cities:

1. The limitation of the area within which factories may be located (like fire lines) and the securing by the municipality of proper facilities for transportation of freight by canal, railroad, subway, etc.
2. The determination by the municipality of the districts or zones within which houses of a given height may be erected, the number of houses which may be erected per acre, the site to be covered and consequently the density of population per acre.
3. The securing by the municipality of the proper means of transportation of the people.
4. The provision of adequate streets, open spaces, parks and playgrounds in anticipation of the needs of a growing community.
5. The right of excess condemnation, i.e., the authority to condemn more than the area to be used for the immediate purposes contemplated by the condemnation, so that ultimately the city makes no net expenditure for land to be used for public purposes.[182]

Marsh's emphasis on transport and openness demonstrates the importance of movement and circulation within the Progressive planning imaginary, while his focus on density and relief of congestion emphasizes the attempt to turn the space of commodity circulation into the place of community. Further, like other Progressives, his advocacy for increased government intervention echoed Progressive sentiment and their "socialist" tendencies.

Marsh's five functions clearly demonstrate the tensions inherent in spatial restructuring during the period. Like other planners, he sees the need to integrate spatial zones or scales into the system of circulation and production, and at the same time wishing to segregate places of community and proper living. Marsh's arguments for increased zoning, improved transport, and proper housing and recreation emphasize the importance placed by planners on resolving the problems and contradictions of space, place, and scale within the context of wider economic processes, and are reminiscent of current debates regarding the global–local nexus and the processes of linking and de-linking discussed in works from Jane Jacobs to Sassia Sasken. With proper planning and imagination, Marsh believes that the local can be made inhabitable for proper citizens, while maintaining the links between the local and the global scales. As Andrew Jonas points out, among contemporary planners, models and metaphors of development at the "local" scale of place are often "an anticipation of the future," an attempt to "map out material scales that eventually might liberate" places of residence and production "from their existing scale constraints."[183] In other words, Marsh's planning functions ensure that the worker's place of residence, adequately provisioned with light, air, and recreation space, is located away from his place of work, but accessible through use of carefully planned transportation paths. At the same time, the worker's place of employment is linked to the larger economic scales of production, distribution, and consumption, ensuring "that factories may be able to compete with others" and not waste manpower and efficiency on needless loading, unloading, and redirecting of raw materials and finished products.

Marsh's five functions constitute a summation of the basic points found in planning documents and papers during the reformscape period. Simultaneous "relief" of congestion and better circulation, the medical metaphors of civic health, constitute the Progressive planner's basic vocabulary of the imagined communities of the modern city. The repetition of such words and phrases like "human suffering," "physical deterioration," and "moral danger"[184] and their imbrication with critiques of bad planning and lack of imagination, constructed a reformscape in which proper visionary imaginings could produce the surgical strike needed to turn the urban problem into the hope of democracy. Architect and planner Charles Lamb combines these elements, including the use of medical metaphors, while envisioning the necessary procedures to create place from space: "For it must not be forgotten that bad planning induces bad building, and bad planning and bad building combined induce dirt and disease, and thus, like the loss of character in the individual, the breaking down of the lines of health and decency go hand in hand with the evils of a bad scheme." Advocating that New York City follow London in its increase in the power of the state to guide and plan restructuring, Lamb anticipates Robert Moses, promoting the use of "the surgeon's knife" to "cut directly through" congested areas. Lamb's lauding of the English model of planning has an almost alchemical quality in the creation of place and community. Referring to the construction of Shaftesbury Avenue in London, Lamb claims the new road, created with the surgeon's knife of state authority and bold planning, achieves dual purposes: connecting two neighborhoods to create a vibrant community and destroying the "plague spot" of the Seven Dials neighborhood. As Lamb states, the process of planned restructuring succeeds, as "Seven Dials disappears from the map of London and, at the same time, from the records of the police court."[185]

Lamb's spatial imaginary, informed by European methods and his own ideas of the city as an integrated work of art, represents the conceptual vision shared by many of his contemporaries. Approaching the restructuring of the American city from a variety of methods and directions, Progressive planners shared certain common assumptions: the need for bold action and government oversight; the importance of open space and easy movement; the integration of the local with the regional; the connection between the creation of community as place and civic pride and responsibility. The combination of these common assumptions framed the conceptual vision of the reformscape, creating a modern understanding of the urban, one in which the organizational logics are contained within the skyscraper and the model tenement, and within the opulent grandeur of stylized buildings and the modest improvement of the small park. The Progressive use of medical metaphors is well established in the history of spatial restructuring; reformers like Felix Adler were well known for their use of language such as "treating the symptoms of the disease" when referring to the urban built environment.[186] What is more obscure is how this rhetoric of malady works with other "planning logics," such as the artistic and aesthetic, to construct "the very language used to evaluate" not only spatial restructuring but the collective understanding of the populations living in urban space. The modern city itself becomes a new mode of knowing. Simultaneously, the process of restructuring brought about by the reformscape imagination opens up the representational space of habitation, producing the heterotopias of spatial redeployment.

The reformscape imaginary, when put into practice, resulted in spatial restructuring in most U.S. cities, including New York. For Hell's Kitchen, restructuring included not just the placement of DeWitt Clinton Park but changes in tenement construction and management, in the form of new laws and the use of various "model" tenement techniques. As well, from 1894 to 1914, improvements were carried out on the waterfront docks, subway lines were completed, electric lighting slowly replaced gaslight, and elevated railways were dismantled, with much of the work and management of restructuring being carried out under the latest in managerial and government techniques, such as the new Port Authority and the reorganized Docks and Sanitation Departments. Reformers like Calvin Tompkins, Lawrence Vellier, and Ernest Flagg were active participants in the restructuring process, serving as expert witnesses before various governing bodies and as agents of city authority. Invoking their ideas concerning the future of the nation and the need for community, their efforts were often aimed directly at aggregate populations, such as the residents of Hell's Kitchen. Their attempt to create place and community utilizing the techniques developed in the urban planning field, while serving to improve the circulation of people, goods, and services, imposed a vision of community sometimes at odds with what already existed, and always at odds with how space and place relate.

KITCHEN SPACE

TO UNDERSTAND COMMUNITY and space in an urban setting, it is necessary to reconsider the fixed, set notions that inhabit the thinking of urban planners, past and present. For reformers working in New York, "community" indicated a notion of society as a "collection of individuals contingently bound" by lifestyle, status, ethnicity, and geographic location. Evocations of community, and attempts to create them, rely upon ideas of human interaction that assume that conflict, difference, and even radical alterity can be properly managed and contained. In the case of Progressive reformers working within the reformscape, attempts to spatially structure a proper community relied upon two overriding assumptions: first, that what previously existed, if it could be considered a community at all, was based upon improper spatial design that promoted negative habits; and second, that an existing, static model of what constituted community could be superimposed on the existing conditions. These assumptions were imposed on forms of cooperation, conflict, difference, and alterity that constituted the existing communities of the Middle West Side, communities that were not static but ever-shifting, responding to and helping to create the very spatial conditions in which they existed.

The conditions in Hell's Kitchen presented problems in determining the shifting levels and scale of local communities.[187] Area residents left little in the way of direct commentary, and that which does survive is mediated by those who collected it: the reformers and social and settlement house workers. Unlike other areas of Manhattan such as Harlem, the Lower East Side, and Yorkville, Hell's Kitchen produced little in the way of "local" dailies or weeklies other than ethnic-based newsletters, which largely concerned events "back home," and its working-class occupants did not produce much written evidence themselves. Determining the shifting levels of community becomes a matter of interpretation, drawn from the evidence that does survive. In doing so, one must determine how space was produced within the framework of the restructuring brought about by the reformscape imaginary. This requires some determination regarding how space was perceived by Hell's Kitchen residents: what were the local meeting points of different groups; where were the popular routes to travel; what new areas were opened up by transportation improvements; where did people go for recreation, for pleasure; what were the priorities and worldviews of different identity groups? Further, what did Hell's Kitchen residents think about model tenements and park space?

The existing, ever-shifting levels of community in Hell's Kitchen in this period of restructuring did not constitute a coherent, unified structure where space was utilized in like manner by all residents and a universal set of norms were enforced. Social solidarities and fragmentations, as well as competing and often contrasting norms and values, were produced along with the mental and physical spaces of daily life. Solidarities often formed along class, ethnic, and racial boundaries, but these same identity forms just as often were fragmented by clashes within the boundaries, as discussed in the previous chapter. Accepted cultural norms, such as the proper spaces for young women, or older men, or the proper space for performing life rituals such as weddings and christenings, would often be matters of dispute among members of groups.[188] In addition, agreement and conflict regarding issues of health, social play and relaxation, criminality, domestic interiors, and gender roles played out within the production of mental and physical space. In order to examine how space structures these processes, it is necessary to look at the local space of "community" to follow the shifting scales of solidarity and fragmentation.

What were the centers of shifting community life in Hell's Kitchen, and what functions did they serve? How did the process of restructuring affect the area, and what did this mean for the amorphous sense of community? Several obvious areas served as points of local activity, some more apparent than others. Certain physical spaces and locations were vestiges of the area's past development, and others represented the efforts of reformscape planners. Paddy's Market, an open-air food-sellers' strip operating on Eighth Avenue between 34th and 38th Streets, and the slaughterhouses near the 44th Street piers are examples of the former, and DeWitt Clinton Park, settlement houses, model tenements like the Emerson Flats, and new recreation piers exemplify the latter. Understanding how restructured spaces were integrated with the vestigial requires some compromise between the place-based analysis of humanist geography and the integrated, holistic approach that views all space as part of a larger, systemic whole. Adopting what critical geographer Nicholas Entrikin calls "in-betweenness," we will examine both new and old spaces in Hell's Kitchen not as an organic whole or as place-based contexts for identity formation, but as something in-between, where the workings of market forces, reformers, and the local population produce both the physical space of daily life and the mental space of heterotopia, which serve both as solidifying and fragmenting elements in the construction of identity.[189]

The production of space in Hell's Kitchen meant different things to different people. Accounts gathered from residents show how categories such as ethnicity, class, race, and gender could bind one to a sense of space, but also fragment those same categories. Ethnicity was obviously important to local residents, and many places of public gathering were based upon ethnic identification and national origin. Churches, social clubs, places of entertainment, and, at times, whole blocks of tenement buildings, were often divided along ethnic lines.[190] In 1900, families with heads of households born outside of the United States totaled some 84 percent of the tenement population. According to the 1900 Census, and compiled by the New York City Tenement House Department, Table 2 breaks down the numbers.

Although Irish ethnicity predominated in Hell's Kitchen, solidarity and loyalty could vary from street to street and parish to parish. For the dominant ethnic group, the Irish, social institutions such as the Catholic Church, Irish social clubs, the working piers (where the Irish dominated longshore employment), certain tenement rows, and the textual space of their ethnic press, were vitally important spaces. Exchanging information, networking, forming political alliances, and socializing were carried out in and around such spaces. Surveys of two local ethnic papers, *The Gaelic American*, a nationalist publication, and *The Catholic Register*, a publication of the local diocese, illustrate the deployment of local space to maintain ethnic solidarity. A location such as Murphy Hall, an Irish clubhouse on 49th Street and Eighth Avenue, was used for regular meetings by the Men's Association of Donegal and of Fermanagh, as well as by ladies' clubs from a variety of Irish counties. Indeed, the club and association meetings, some held on the West Side, some farther east,[191] highlight the Irish community's relationship to events in Ireland, at both the level of national politics and of local happenings, both political and social. Much of the activity of these county clubs went into fund-raising and maintaining public awareness of what the *Gaelic American* termed, in a weekly segment, "the work at home," supporting efforts in Irish counties to teach Irish history and the Gaelic language. Efforts such as supporting the "work at home" served to solidify community ethnic feeling, maintaining a

Housing and Visible Spaces

Table 2: Tenement House Report, Country of Origin

	Number	Percent
Ireland	8,352	37.2
Germany	6,649	29.7
United States	3,619	16.1
England/Wales	784	3.5
Italy	578	2.6
Scotland	413	1.8
Nordic	259	1.2
Russia	172	1.2
Other	1,124	5.8

connection to a cultural, ethnic heritage that is often stronger in emigrant communities than it is "back home."[192]

For the West Side Irish, connections to Ireland, in the form of relatives and cultural memory, provided the framework for the production of local spaces of ethnic identity. As Ireland experienced numerous instances of struggle with England in the late nineteenth and early twentieth centuries, the cause of Home Rule and anti-English settlement served as rallying points to unite ideologically opposed groups. Whereas larger organizations such as Clan na Gael were primarily the reserve of the more well-off members of the New York Irish diaspora, county clubs, such as the Donegal Men's Club and its Ladies Auxilliary, which met in Hell's Kitchen, were largely the preserve of the Irish working class.[193] County clubs created spaces not only for political activity, as when the clubs united to oppose the peaceful mediation of U.S.–British disputes in 1896, but were also places where business connections could be made. Though the majority of members were drawn from the working classes, county clubs also attracted a good number of Irish professionals and small business owners, and became places where a good carpenter could connect with a local contractor. The county clubs further contributed to the expansion of Irish ethnic space through participation in sports. As Manhattan clubs grew in size, they pooled their resources in 1897 to purchase land in the Laurel Hill section of Queens. There they built Celtic Park, originally set up to provide grounds for the traditional Irish sports of curling and Irish football. Only a ten-minute train ride from the Manhattan ferry terminal in Long

Island City, Celtic Park's surrounding environs quickly became an Irish enclave, housing working-class and professional Irish leaving Manhattan in search of better living conditions.[194]

The Irish ethnic press was also utilized as a weapon to defend the Irish New York community against racist slurs, and to identify political friend and foe. William Randolph Hearst and his paper, the *Evening Journal*, came in for particular attention and vitriol in the *Gaelic American* for "slurs" against Irish women, and reformers like Dr. Charles Parkhurst were often singled out for their characterizations of the Irish as monolithic, despised, and inebriated. The *New York Sun*, the "recognized organ of the English in New York, although owned by a Jew," appeared as a weekly foil in the *Irish American*, and its bias was answered swiftly and decisively. Even fellow Catholics were called on the carpet, as when the *Gaelic American* conducted an ongoing campaign between 1900 and 1905 to end the "caricaturing of the Irish" at Catholic street fairs. But the Irish ethnic press did not only defend ethnicity. It promoted the Democratic Party in local elections, objecting especially to the "tribute" paid by Irish city dwellers to upstate interests. Irish trade union papers were particularly apt to promote homeownership, often framing the issue in class terms, defending the right of the "trades man" to own his own home.

Churches were obvious places of ethnic and religious solidarity, as well as meeting places and nodal points for many residents. For the Irish, St. Malachy's on Eleventh Avenue and 40th Street, and St. Raphael's, also on 40th Street, dominated community worship.[195] Italians in the area traveled to St. Michael's on 34th Street and Seventh Avenue, or to Holy Cross, at 42nd and Eighth. Local African American families tended to gather at St. Benedict the Moor at West 53rd Street, whose congregation would follow other blacks to Harlem in the 1920s. Demonstrating the changing demographic patterns of the area, the first Catholic Croatian church was established on West 40th Street in 1913, in a building that had previously hosted Irish, Italian, and Polish congregations. German residents could attend St. Luke's Lutheran on 46th Street between Eighth and Ninth Avenues, a German-Catholic Church of the Assumption on 49th and Tenth Avenue, or the German Baptist church on 46th Street, whose site would later become the Wessell, Nickel and Gross Piano factory.[196]

Religious life centered on the church parish also provided a framework for the production of gendered spaces. In her research of church fairs in Manhattan, historian Colleen McDannell shows how the fairs became spaces dominated by women, who took leading roles in organizing, setting up, and running the events. Church fairs, held to raise money for worthy causes through the selling of donated goods, were not only forms of entertainment but allowed spaces for women to actively participate in the public sphere of the city. As McDannell documents, though the men of the local parish performed some of the labor it was the women who "collected donated items, created festive environments, staffed the tables, cooked, and supervised the whole affair." Parish church schools also served as spaces of resistance to the dominant Protestant culture of public education.[197]

The waterfront dominated the spatial environment of Hell's Kitchen, providing space for economic activity, employment, and recreation. The waterfront area was also the site of much of what gave the district its reputation. Though one pier at 50th Street had been turned, by 1900, into a recreation pier, its immediate neighbor to the south was the location of ash dumping, where a near constant-stream of horse-drawn, and later motorized trucks dumped Manhattan's coal ashes in an endless procession. Eleventh to Twelfth Avenues north of 34th Street were dominated by the rail yards of the Pennsylvania and West Shore railroads. Stock pens and slaughterhouses spanned the same area from 40th to 42nd Streets.[198] Despite the unsanitary conditions of the North River and the dangers associated with waterfront locations, the piers of Hell's Kitchen were a primary gathering point for groups of young men, who swam in the river in summer, picked up odd jobs hauling, and often used the piers as unofficial "clubhouses" for black market activity.

For tenement dwellers, whose spatial interiors were confining, the streets of Hell's Kitchen were a primary space of interaction. Front stoops and steps were places for informal gatherings, where men often shared a beer before turning in, or young men and women could socialize within observing distance of the family. It was in these public spaces that social norms and values were formed and contested. Issues such as who was deserving of city and private charity, when it was proper to steal, proper and improper behavior for young men and women, and the relative merit of local politicians were all issues worked out in the improvised public sphere of the street and neighborhood. In his reminiscences of childhood in Hell's Kitchen in the early 1900s, former resident Tom McConnon recalls the lessons learned from ice lines: "The free ice line provided many revelations of human nature. There were those who were poor and were not too proud to make the world aware of it.... There were the poor who were ashamed to let it be known . . . the outright chiselers," those who could afford to purchase their own ice, but "wouldn't miss an opportunity to get something for nothing." One lesson, born of the experience of the economic insecurity that constantly buffeted residents, was never to put on airs during flush times and never to look down on the deserving charity cases. McConnon describes the particular disdain residents held for those who lied to investigators from the Society for the Improvement of the Condition of the Poor to gain their "blue tickets" for the summer ice giveaways, and remarking on the dignity with which truly needy families carried themselves in the very public distribution of this form of charity.[199] He also recounts the tale of Daisy, an "actress," who, in flush times had purposefully made her purchase of the day's ice from a local ice-monger, making a public show of her ability to pay. Later, when she is forced to join the charity line, barely concealed conversations speculate over her source of former income, which many attribute darkly to "her dozens of men friends" who "take care of her," the inference quite clear.

McConnon's memoirs contain similar testimony regarding property laws and appropriations like those contained in social work interviews with Hell's Kitchen residents, a moral relativism that offended social workers and caused them to proclaim residents to be devoid of values. When McConnon's family occupies their new apartment in Hell's Kitchen, his sister demonstrates no moral qualms about appropriating paint left in an empty apartment on the floor below. Her reasoning that a new paint job is what the family "deserves" trumps any objections her religious mother has to stealing. Appropriating coal, ice, old fruit and vegetables, wood scraps, and plumbing fixtures was a common practice among young people in the area.[200]

Rooftops in the tenement districts were special places of socializing. Pigeon keeping and kite flying were two important social activities that took place on the roofs of Hell's Kitchen. McConnon

Housing and Visible Spaces 147

The rooftop playground of Hell's Kitchen youths. The airshaft is an example of "old law" dumbbell tenement style. (http://www.archives.gov/)

recalls the kite types, from the homemade paper bag on a string to the "two-center" purchased at the candy shop, the relatively unaffordable box kite, and the "Chinese kite" made for his seventh birthday by a local blacksmith. Both adults and children engage in the activity, which provides amusement and competition among the neighborhood buildings. According to McConnon, skill in kite flying is exceeded only by skill in "kite slinging," the practice of pirating the kite of a nearby rooftop by attaching a heavy object to a string and "capturing" or pulling in the lead line of a well-constructed kite. According to McConnon, the dramatic battle over the attempted capture of a kite broke down ethnic solidarity, as one rooted by building in such cases.[201]

The tenements themselves, a mix of pre-reform and newer structures, provided spaces of interaction and contestation. Though new laws governing the construction of tenements were passed in 1901, "old law" buildings based on the 1879 law, the infamous dumbbell tenements, so called because their blueprints resembled dumbbells with narrow airshafts cut between adjacent buildings to comply with the laws requirement for ventilation, continued to dominate the area. Tenement space, including the alleys, hallways, rooftops, steps, and interiors, provided sites of interaction between tenants of different racial, ethnic, and religious groups, as well as between tenants and landlords, public authorities, and social reformers. In addition, the preference of some residents for model tenements, and the avoidance of the models by others, indicate differences in what was valued among members of the Hell's Kitchen community, values that often shifted over time as space was restructured. While some residents clearly preferred the orderly yet paternalistic atmosphere of model buildings, others chose to avoid the rules, regulations, and screening process such buildings required of tenants.

Churches, ethnic clubs, the waterfront, outdoor markets, the streets, and tenements were spaces where daily life was performed by residents of Hell's Kitchen. They made the physical space of the Middle West Side into a community whose values and norms, like most communities, were neither rigid nor static, but contested and negotiated. All were affected by reform plans and efforts to restructure the physical space of the area. In order to gauge how restructuring changed the area, and how local residents produced space, I will examine the impact of Clinton Park, new tenements, and the changing landscape of public space in Hell's Kitchen.

DeWitt Clinton Park officially opened for full public use at a ceremony attended by city political luminaries such as New York City's corporation counsel John J. Delaney and Parks Department president Samuel Parsons, on November 5, 1905. According to press reports, some 5,000 people were in attendance for the ceremony and opening activities, which included races, basketball, and other games. A *New York Times* reporter claimed the park "one of the most beautiful in the city," due to its location on the "sloping ground toward the North [Hudson] River." The 7.4-acre park, designed by Mr. Parsons, featured a recreation-bathing pavilion, gymnasium, running tracks, playgrounds, and a series of curving paths

The Children's Garden in Dewitt Clinton Park, 1905. (http://www.loc.gov/)

with panoramic views of the North River and Palisades. Of course, as mentioned, the park had been in use since 1902, when Frances Parsons [no relation to Samuel], the city's first woman park official, had organized the 'children's garden' to teach local children about "plant science, conservation, nutrition, and hygiene."[202]

The children's garden had become something of a sensation among reformers and local philanthropists. Officially designated the Farm Branch of the National Plant Club, the garden was run by six administrators and divided, by 1903, into 432 separate lots, containing beets, radishes, beans, carrots, onions, lettuce, and corn. The garden was visited by local teachers, reformers from Europe, and even approved by a delegation from the National Playground Congress. For Mrs. Parsons and her charges, the use of the garden was to be "solely educational," and not intended as a sociological experiment or for the "betterment of any particular class." Mrs. Parsons wished to reach the "children of the brownstone fronts" as well as those of the tenements. Along with giving lessons in basic gardening, Parsons and head administrator Olsen organized exhibits designed to demonstrate the need for proper nutritional balance, and for natural preservation.[203] In one exhibit, an 8- by 8-foot plot was constructed to demonstrate the ill effects of erosion caused by deforestation.

Though the park and the children's garden brought praise on their creators, and were used by many Hell's Kitchen residents for a variety of purposes, the park was not without problems. Whereas Mrs. Parsons remained committed to an ecumenical and educational use of the garden, her supporters in the powerful Citizens' Union[204] clearly saw the intended audience of her efforts as the recalcitrant tenement dwellers who so alarmed them. In a letter praising Mrs. Parsons's efforts, and offering financial support, Citizens' Union president R. Fulton Cutting saw the benefits of such an endeavor: "When the city can see the way to the formal incorporation of a movement such as yours we shall not need to spend so much money in the preservation of order and the punishment of crime. Nothing is more refreshing in a great city than such efforts to overcome evil with good."[205] That is, the establishment of park space was a way to ensure the assimilation of what Cutting viewed as a criminal population, which was far more typical of the period, and of reformers, than Mrs. Parsons's enlightened, if somewhat naïve, attempt to be purely educational. The Parks Department official report reflects this attitude, stating that the efforts of Mrs. Parsons produce more appreciation of the city, and less vandalism.[206]

The space of DeWitt Clinton Park, the result of the reformscape vision of bringing place to space, faced other difficulties as well, mainly in its immediate and long-term use. Even Mrs. Parsons's Children's Garden encountered difficulty in holding its intended users. Each plot was assigned to a group of children, with a post naming those responsible for each plot. By 1906, attendance problems, probably brought on by the multiple responsibilities and challenges that young people faced, meant that many plots were poorly tended. Use of the plots was considered a privilege, and could be revoked for infractions such as poor attendance, bad behavior, or loss of ID tag. One user was heard to remark that the name posts, used to designate each child's small plot, would be better utilized as baseball bats. Though the farm plot and park drew praise from city officials and outsiders, several accounts bring into question how the park was viewed and used by those in the immediate area. Pauline Goldmark reports, in *Boyhood and Lawlessness*, that the park was often "practically empty" on a "fine Saturday afternoon."[207]

The park area, according to local informants, had previously been known as "the Lane," a last vestige of the old farmland running down to the river. It had been a small strip of wooded, overgrown land, with a worn footpath, used by children for play and couples for "strolling" and perhaps other activities. The emptiness on a fine Saturday is, at least in part, attributed by Goldmark to the park's sterility, its very ordered appearance. In place of the lane, Goldmark states that "the usual restrained and well-kept air of a small city park is very noticeable," with "uninteresting" concrete paths leading to "tall iron fences" that "perhaps necessarily" enclose the running tracks and courts. Goldmark muses that in planning the park to conform to the landscape and follow the design of a fine small park, "an opportunity was lost" somewhere in the park's construction. From her description, the organic use of the area of "the Lane" was converted into the controlled and regulated space of reform when consultation with potential users might have yielded better results.

The park did serve several purposes, both casual and official, and did become, by some accounts, a nodal point for the immediate surrounding streets. The pavilion at the park's center, known as the "White House," was used for weekend dances attended mainly by the younger crowd. It was the sight of annual events, such as the city-sponsored Fourth of July Athletic Carnival, where races, relays, and other displays of athletic skills by local youths were presided over by referees drawn from the area's professional and amateur athletic circles. Many residents in their reminiscences recall utilizing DeWitt Clinton, but with no particular fondness. Cathy Yelenock, born in the district in 1901, remembers dances at the White House and tending one of the children's gardens, but claimed the best dances were at the Hartley House settlement on West 46th Street.[208]

Efforts after 1900 to continue park construction for the Middle West Side area proved futile, and the long-term prospects for DeWitt Clinton Park were not promising. Ruth True describes the park, in 1914, as "small, restricted, and inadequate." No other major parks were built in the area, and DeWitt Clinton Park eventually lost part of its territory to the West Side Highway, clearly a more important priority for the city as a whole. In 1908, concerned citizens, led by local public school principal Henry G. Schneider, organized the West Side Children's Playground and Recreation Conference to advocate for more city parks and for organized forms of activity in the area they dubbed "the neglected half-mile square."[209] But despite their lobbying efforts, no additional park space was added to the Hell's Kitchen district. Schneider and his fellow members blamed political patronage and corruption, and favoritism shown to solid Tammany districts of the East Side for the ongoing neglect. Despite consistent appeals for investigations into conditions and for new park construction to, as Schneider put it, "give the voteless babies a chance for life," the construction of DeWitt Clinton Park was the only major achievement of park reformers in the area.

Model tenements and new tenement laws were other reformscape efforts to provide proper spatial provision for urban residents.[210] Like parks, new tenements were designed for everyday use, and were the result of a common vision concerning space and behavior. Two new types of tenement were part of Hell's Kitchen's restructuring: the model tenement and the "new law" buildings designed to allow more light and air. Both were meant to provide proper space, more open and visible, to urban residents. But the struggles over the physical construction of tenements and the rate of profit to be gained from rentals were only part of the spatial story. Tenement living was, of course, represented in numerous ways as a main part of the urban problem, and those who resided in tenements were, by reformscape definition, part of the problem. Whether a family was of the "deserving poor" or considered "shiftless" and "lazy," the fact of renting in crowded buildings without the privacy afforded by the detached "home" and without "protection" from unsavory elements positioned tenement dwellers as objects of study. But tenement dwellers in Hell's Kitchen often waged their own battles—among themselves and with public authorities and public opinion—over rules and social norms.

The reformscape restructuring of the urban environment centered, in large part, around the home space of the urban poor. Two major projects, one to improve the physical construction of tenements and the other to "involve" tenants in the daily upkeep and keep rents affordable, spearheaded the efforts. In the former, changes in laws governing the building of tenements were passed in 1861, 1873, 1890, and 1901. By the time of the 1901 New Tenement law, "Old Law" buildings constructed in the dumbbell style had been banned from future construction, but New Law structures still lacked the provision of space demanded by housing reformers like Lawrence Vellier. Tenement law changes involved city regulation, and were largely the work of urban reformers contracted by the city to form special commissions and conduct design competitions. The changes in laws and enforcement brought about through these efforts were real and lasting, but of limited effect for areas like Hell's Kitchen. An investigator for the Bureau of Social Betterment found in 1914 that only two buildings in the four-block radius around Ninth Avenue and 40th Street were constructed under the new tenement laws. Many of the buildings that remain today in the Hell's Kitchen area are Old Law tenements. No provision in the new statutes passed after 1901 called for the destruction of old tenements, nor was any government funding forthcoming for such a project. In fact, a survey of building dates along Tenth Avenue between 48th and 52nd Streets shows that many tenements were constructed in 1900 and 1901, indicating a building rush to beat any new laws. Because of the complexity of the new tenement legislation's regulations governing construction, the cost of erecting new tenements required greater capital investment and the hiring of professional architects. For the Middle West Side, construction of New Law tenements did not really begin in earnest until the postwar housing boom of the 1920s.[211] As a result, most of the buildings housing Hell's Kitchen residents were and remained inadequate.

The attempts to build and operate "model" tenements, on the other hand, were largely private efforts carried out by urban philanthropists and reformers. The guiding idea was to find a group of investors who were willing to settle for less return than would normally be paid by a market endeavor. Though model housing numbers were small compared to ordinary tenements, the effort received widespread attention from the press and city officials, and was a main area of focus for reformers like Felix Adler and Ernest Flagg, as well as philanthropists like oil magnate Charles Pratt. As Plunz points out, many early advocates of model tenements drew on English models from the 1850s to 1870s. Felix Adler's Tenement House Building Company was formed in 1885 with money raised from a board of directors who agreed to limit their profits to 4 percent. The architectural plan called for two- and three-room apartments and community space on the ground floors in a building of six stories. Adler also insisted that owners take an active interest in the daily operation of the buildings to combat the "great evil" of absentee landlords.[212] The company's only successful construction was on Cherry Street on the Lower East Side. A far greater and more successful model tenement effort was undertaken by the City and Suburban Homes Company, established through the efforts of the Improved Housing Council and the Association for Improving the Condition of the Poor in 1896. Backed by Adler and E. R. L. Gould, and utilizing building designs by Ernest Flagg, City and Suburban became the largest builder of model tenements in the United States.

Though many of City and Suburban's major projects were constructed on the East Side, the Middle West Side was the recipient of several projects, including the Flagg-designed Fireproof Tenement Association buildings on Tenth Avenue between 41st and 42nd Street, and on West 47th Street. City and Suburban also financed the construction of model tenements for "colored" tenants on West 62nd and 63rd Streets. Model tenements were promoted through exhibitions staged by housing reform groups like the House Committee of the Charity Organization Society, whose 1898 exhibit included scale models of city blocks with "Old Law" buildings next to blocks of model tenements, charts and graphs illustrating density figures, cost of materials, rentable space, and photographs of existing model buildings. With the recognition that market forces and private profit would always mediate attempts to improve tenement law and regulation, model tenement construction became the preferred method of reformers for providing the so-called deserving poor with proper space.

The attempt to regulate tenement construction and to operate model tenements that provided affordability and community involvement met with mixed results and ambiguous receptions. The new construction laws did not significantly alter the crowded conditions of most tenements, nor did they provide the light and space demanded by reformers. Model tenements often failed as business enterprises as market pressures forced changes to the original designs, and many Hell's Kitchen residents rejected the paternalism and rules of the new operations. Reformer Elsa Herzfeld, in a bit of obvious hyperbole, reports in 1905 that her year-long investigation of tenement dwellers on the Middle West Side revealed that they "hate the model tenements, which have been built for their especial benefit." She attributes this "curious antipathy" to the "instincts" of those living in the area, who are anti-authoritarian and resistant to notions of "good citizenship."[213] But the reasons for not wanting to live in model buildings are repeated in other testimonies, and their similarities reveal an anti-paternalist sentiment felt by many in Hell's

Kitchen. Miss Herzfeld lists the objections to "them new-fangled homes" as higher rents, intrusive house wardens, social pressure to maintain standards of cleanliness, and prohibitions against use of fire escapes for storage, socializing, and clothes drying.

Of course, not all Hell's Kitchen residents rejected model tenements. A 1913 report of two new five-story model tenements built on West 44th Street near Tenth Avenue suggested that all units in both buildings were rented immediately upon completion. The new buildings, constructed with money invested by Catherine and John Rogers, were cheaper than other model tenements at five dollars per room, and offered amenities such as electric light in each unit, hardwood floors, marble-floored hallways, and a basement garbage incinerator that produced heat sent back to each unit. The building also included a rooftop children's play area, lines for laundry, and a second-floor reading and social space, equipped with bookshelves, high ceilings, and plush chairs. The basement and ground floors were also equipped with "perambulator rooms" for those with infants or small children. It was also reported that the Rogers couple had placed strict restrictions on the residents, and that the application process was described by one local as requiring a "regular civil service to get in." Mr. Rogers, a member of the Knickerbocker, Tuxedo, and Downtown clubs, confirmed that a "careful search into the history and character" of all potential tenants was carried out. Though the buildings provided amenities not otherwise available, they also produced the paternalism and oversight that many residents resented and rejected.[214] Restrictive background checks led to charges that many model tenements in Hell's Kitchen and other parts of the city were occupied not by those E. R. L. Gould designated "the worthy poor," but were snatched up by low-paid white-collar professionals and well-paid workers who could withstand the scrutiny of a background check.[215] The excessive delegation of space for "community use" was also a major problem model tenements encountered in keeping profits at the minimum required to sustain investor interest.[216]

Responding to critics like the architect Henry Atterbury Smith that "no model tenement has changed for the better the living conditions of the very poor," E. R. L. Gould claimed that model tenements neither housed the middle classes nor were inherently unprofitable. Gould claimed that model tenement residents would naturally be drawn from those among the working poor with "ambition" and a desire to improve. Those among the "distressingly poor," the "incompetents and derelicts," would likewise "naturally reject" the "maximum standards of law" that prevailed in the new projects.[217] For Gould and other housing reformers, model tenements were learning tools, teaching the working classes the value of property and providing the proper space for "uplift." Some went so far as to claim that "the ultimate fate of the Republic" depended upon providing the proper spatial environment for the urban poor and their children. Gould was answering charges that model tenements often were rented by those who could afford better housing in the open market while those who "deserved" the chance at affordable living space were often crowded out by both the screening process for residency and the competition for the apartments. Model tenements faced a variety of problems, from increasing pressure to maintain profits while renting below market to restrictions placed on safety and building materials that increased construction costs. With the housing crisis brought on by the U.S. entry into the First World War, model tenements were a threatened species of proper space.

As many residents rejected model tenements, and some embraced them while they lasted, others put them to use producing spaces that differed from the effects hoped for by reformers. Women in particular utilized the opportunities provided by model tenements to produce gendered spaces. They were often preferred by organizations such as City and Suburban Homes as managers of the model buildings. Women managers were often selected from the resident pool, as E. R. L. Gould and others believed that their "organic" connection to the tenants and "natural" disposition made them superior for settling disputes and seeing to the health and well-being of their charges. Gould further stated his preference for women as rent collectors: "ladies of education and refinement" whose influence is "uniformly good." Most model home managers also expressed a desire for single, self-sufficient young women as tenants, believing them to be "conscientious, frugal and saving."[218] As a result, model tenements offered an opportunity for specific performances of gender roles that expressed alterity to the normative conditions of the larger community.

Two model buildings, the Emerson, or as it was known locally, the Flats, and the Windmere, provided opportunities for West Side women to explore gender performance. The Emerson, built across Eleventh Avenue from DeWitt Clinton Park in 1905, provided a particular and unique spatial environment until its conversion into a more typical, profitable tenement by 1910. For a brief period, the Emerson's first floor was given over to a baby nursery, a storeroom for baby carriages, play space, and a recreation room for reading and quiet. The basement provided shower and bathing facilities, and the rooftop was given over to children's play space and a small garden. For a time, this female-dominated space became the hub of neighborhood activity, attracting not only tenants but also women from the neighborhood, who used friendships and acquaintance with tenants to take advantage of the Flats' spaces. Located near the Stryker's Lane Community Center, and across from DeWitt Clinton, the Emerson's first floor, entranceway, and sidewalk became a gathering space for women, and where it was unacceptable for men to gather and drink. Here, women watched each other's children, socialized, shared tips on affordable food and clothing, and passed information about job opportunities and women's social gatherings. The extension of family space from the private apartment to the first-floor common areas, and the provisions of day care, cooking classes, and literacy programs, opened up space for tenants not previously available.[219] Although the Emerson's ownership group would succumb to market pressures and turn the first floor into rental space by 1910, the brief period under the original conditions gave many tenants an experience of daily living that differed from the conditions of the Old Law tenements many had arrived from. The creative redeployment of the space provided residents with an altered vision of the domestic interior and of how space could be reimagined to meet their daily needs.

The more famous Windmere, on Ninth Avenue at 57th Street, was constructed in 1884 as a model apartment house for middle- and lower-income families. Under the management of Henry Stirling Goodale, an artistically inclined nonconformist, the building became widely known as a haven for "bohemian" women. But the definition of what constituted a "bohemian" woman, and who actually used the Windmere, tell a different story. An 1898 profile in the *New York Times* sensationalized the Windmere as a home for the "New Woman," and indeed some woman tenants were drawn from the city's artistic and alternative community.[220] But studies of tenant records also show that many of the young single women living at the Windmere between 1890 and 1920 were of modest means and aspirations, drawn from the city's army of seamstresses, housemaids, factory workers, and cleaning women, and chose to live at the Windmere for a variety of reasons.[221] For many, the opportunity to escape from their family homes would have influenced their decision. For others who expressed the desire to live in a female-dominated environment, proximity to new erotic potentials may have been a deciding factor. For whatever reason, by 1900, 80 percent of the Windmere's 200 tenants were single women below the age of forty.[222] By dividing the three-bedroom apartments, originally meant for families, into smaller units, Windmere proprietor Goodale created the type of space preferred by young working women whose wages could support an independent lifestyle. As social worker Esther Packard reported after interviewing young women in the Windmere and other "socialized" flats in model tenements, in some units, two or three women shared a bath and kitchen and common room, and in others, small one-bedroom units with private baths allowed for more seclusion. Packard reports that by 1916, many women employed by local businesses and industries preferred buildings such as the Windmere to furnished rooms in "women's homes" that advertised themselves specifically for young working women. These homes were normally sparsely and "tastelessly" furnished, with poor-quality food and restrictive rules enforced by matrons who tended to "intemperance and uncleanliness."[223]

Young women who resided at the Windmere were challenging the norms regarding single women and apartment life. Middle West Side mothers, in particular, worried about the "reputations" their daughters might attract if they lived in apartment houses known to cater to young, single women.[224] Many of the young women, according to Packard's report, objected to the lifestyle restrictions imposed in their family homes as well as in furnished rooms, which were often subsidized by charity organizations. They preferred the independence, privacy, and camaraderie of private apartments and socialized flats shared by groups to the "gossip, charity, rules and regulations" of organized women's homes.[225] Many of the women Packard interviewed expressed their desire for privacy and for social space, both to entertain guests and socialize with other women.[226] The threat to existing dominant norms is clear, as when the *New*

An image of the Windmere Model Apartments, as it looked in the 1990s. (http://forgotten-ny.com/)

York Times story on the women at the Windmere presents them as "new women," whose private lives are "not necessary to go into" with implications of illicit or immoral activity.

Tenement interiors were important as produced spaces in Hell's Kitchen. For families especially, lack of interior space forced compromises in living conditions, and pushed normally private behaviors into public space, blurring the boundary distinctions so vital to the model middle-class public sphere. Thus the proper roles of men and women, and proper behavior of children and young adults, were contentious issues worked out often in the liminal space between the public and private. Social worker Elsa Herzfeld reports the contempt male household heads felt for women's clubs promoted and organized by social settlement house workers and the insistence of the household's women on their right to attend. Most men seemed to consider the meetings a lark, where women dressed in their best clothes and frivolously wasted time better spent on domestic work, whereas women saw the meetings as important social and economic gatherings where information was exchanged regarding household survival strategies. Arguments and disputes concerning women's attendance at such meetings often began in the home's kitchen, then spilled over into the hallway, and were continued in men's and women's social spaces, such as the bars and sidewalks.[227]

Issues such as the proper places and roles of male and female family members were points of contention that reflected the physical conditions of the Middle West Side and the changing norms and mores of the larger society. Herzfeld's work on twenty-four Middle West Side families in 1905 demonstrates some of the tensions pulling at Hell's Kitchen residents, particularly women. Tension existed between traditional methods of healing and medicine, such as salves and powders, and the increasingly scientific emphasis on germs and communicable disease promoted by health professionals. Issues around health and sanitation also produced contention between West Side residents and social service providers, who were often at odds over basic questions of child rearing and cleanliness. Combined with the need to keep children home, for example, to care for ill siblings while both parents were working, disputes between parents and social workers were constant sources of tension.[228]

In Hell's Kitchen, like other urban neighborhoods, public and private space was a matter of contention, and its use produced not the "place" of community, but a multitude of ideas about how space would be used and deployed. Small battles were waged over churches and streets, between ethnic groups and within families; over where to shop and purchase by labor leaders; over proper behavior and social norms; and over kite flying on rooftops. Herzfeld reports that a main point of contention among married couples was over church attendance, with women generally favoring regular attendance and men attending reluctantly, if at all.[229] Larger battles would be waged in public spaces of city government, over removing train tracks, demanding city services, and access to political power. Large and small struggles were waged over questions of belonging, ethnic pride, class bias, and gender roles. In public space, both the physical space of the storefront sidewalk, the park, and saloon, and in the space of public opinion, solidarities and fragmentations formed and re-formed. In private spaces of residence, certain families and individuals sought to emulate middle-class norms and conventions, while others rejected or modified these templates. In public spaces, such as "Devery Park," unofficially named for Tammany leader William Devery, political issues such as reform were played out in often raucous circumstances.[230]

As economic development and public provision increased opportunities for mobility, struggles over values and norms played out in the restructured public spaces of Hell's Kitchen. For certain residents, the heterotopic space created by restructuring might produce the potential for mobility in housing or employment, opening up new avenues to different conceptions of proper space and lifestyle. For others, the failure of so many reform efforts, the persistence of inadequate housing, the lack of regulation of the waterfront, might have opened up a different view, a heterotopic space of horizontal movement that kept them bound to the area, both physically and mentally. The enticements of real estate salesmen and advertising promoting the American Dream of single-family suburban home-ownership would entice some Hell's Kitchen families to leave the area, but the lack of opportunity and resources caused many more to remain.

The well-meaning attempts of reformscape urban Progressives to create the spaces of community they deemed necessary for proper citizenship failed. Community could not be created, nor place made, by proper planning and provision. The failure was not the result of misguided action but of conjoined economic, political, and social factors that determine the actual processes of constructing and restructuring urban space. The reformscape vision of community and place was reliant upon the empty spaces envisioned through utopian conceptions of the urban that were not feasible in the built environment of an existing industrial city.[231] For the reformscape vision to become the everyday reality of lived urban space, all factions within the city would have had to come to some consensus over issues of housing, transport, allocation of resources, and most important, profit and accumulation. Though reformers tried to form the rational public sphere that would be receptive to their vision, their overall project, the creation of community based on rational distribution of resources, failed to transform the public into the monolithic block required for such work. What they did succeed at was the implementation of limited spatial changes, and the creation of epistemological categories regarding urban problems and their potential solutions. Both results had deep and lasting effects.[232]

But the real failure resulted from the narrow outlook of reformers who could not see that community already existed on the Middle West Side, albeit not in the form they envisioned. Community, in the reformscape sense, could only take place when the proper provision of surplus profit, public services, spatial environment, and individual self-discipline came together to form self-governing communities of citizens who eschewed the politics of class, race, or ethnicity for a politics of the public good. But as Warren Magnusson points out—and he is worth quoting at length—the practice of self-government and community formation involves experts, officials, and government "only indirectly. People sort themselves out at bus stops, on sidewalks, in restaurants, on waterways and in parks. These intimate practices of regulation and self-regulation involve plays of power, shows of authority, threats of violence, calls to solidarity, habits of deference, and challenges to the existing order."[233]

Just as it is not possible for community to be created, it is not realistic to understand the subjective meaning of space and spatial environment for residents of any given area. Yet it is possible to try to determine which representational or heterotopic spaces were produced by reformscape spatial restructuring by examining how the changes in the built environment were received and utilized by residents. The tension between "Fifth Avenue and Tenth Avenue," to be discussed further in chapter 6, undoubtedly played a role in spatial conceptions. Spatial restructuring in New York City between 1894 and 1914 did contribute to changes in the way people lived, but did not create a rational city of proper citizens. Instead it produced spatial conditions containing multiple potentialities that often reproduced existing social relationships as they were contained within the powerful discourse of community and place. And yet, restructuring also opens up potentialities that escape immediate containment.

What was produced was new physical space within which local residents could construct and reconstruct their daily relationships, with each other, with landlords, with city authorities, and with the wider world around them. These new physical spaces, such as new tenements, subway lines, and parks, along with improvements in city services, opened up for Hell's Kitchen residents new opportunities for mobility, both physical and cognitive. As Foucault's heterotopias are real spaces that "juxtapose several sites that are incompatible," these new spatial environs opened for urban workers representational spaces that created new wants and desires, though these were also multiple and often incompatible. The multiple potentials for active production defy and exceed the ability of city planners and reformers to anticipate the probabilities. As the daily paths traveled by Hell's Kitchen residents were altered, and new spaces became landmarks around which people gathered and physically navigated the built environment, new space combined with existing structures to alter the vision not only of the present, but the future as well.

In negotiating the daily rounds within the built space of the Middle West Side, Hell's Kitchen residents perceived the changes in the physical structures not as "place," as many urban planners would prefer, but as an active space of production where their own perceptions of space contributed in important ways to the processes of production. Though most spatial processes of restructuring did tend to reproduce existing social relationships, the collective perceptions of residents did on occasion create heterotopias, the non-material spaces of production where incremental change "takes place." As the reformscape imaginary attempted to produce the places of community that they felt Hell's Kitchen residents lacked, the new model tenements, small parks, settlement houses, and baths served to accumulate knowledge for urban reformers, at the same time capturing and redirecting the temporalities of those living in these imagined spatial communities. Community in this sense was imagined on both ends. Reformers imagined that the spatial provisions of rational restructuring would create the place of solid, respectable community. Residents of Hell's Kitchen, like residents of other underserved urban areas, reimagined their community on a daily or regular basis. What takes place is not the creation of community or the security of knowing one's surroundings. Rather, this urban space is constantly produced in relation to other spaces, the many-faceted multiple nowheres of heterotopia.

CHAPTER FIVE
SPATIAL ECONOMIES

*Sometimes I throw up a job on the first day. I can tell.
I look around and see that it's not for me. Then I
work out the day and go back home.*
—HELL'S KITCHEN RESIDENT, 1914

*Instead of one girl at each loom, four girls can work eight looms,
with a fifth watching for mistakes and filling in.*
—HIGGINS CARPET SUPERVISOR, 1898

ON OCTOBER 8, 1900, a spokesman for the E. S. Higgins Carpet Company made public what 2,000 workers in Manhattan had feared for some time. Higgins Carpet, a fixture in Hell's Kitchen since the 1870s, a major employer of both skilled and unskilled labor, was moving its production facilities from the current location on Eleventh Avenue and 44th Street. The stated reason for the relocation, according to the Higgins representative, was the high tax rates and water bills the company paid at its present location.[234] Higgins's announcement of its intention to move brought on a slew of rumors, reported in the business press, of other locales that were attempting to entice the company to relocate. Many of the rumors concerned southern U.S. cities such as New Orleans and Atlanta, where manufacturing was booming due in large part to cheap labor, and a "friendly" business environment. As city boosters competed to lure new businesses, Chambers of Commerce were eager to attract companies with tax breaks and the promise of low-wage labor peace. The carpet manufacturer entertained offers from southern cities even though internal documents show that Higgins executives were aware that supply networks and other factors limited their relocation choices to a 250-mile radius.[235]

In a period of local, regional, and global competition, the spatial fix sought by the Higgins corporate board is not surprising. Small to midsize manufacturing firms like Higgins were under constant pressure to maximize profits in an era when consumer choice and transportation improvements meant that competition came from all corners of the globe. Higgins would eventually relocate to Connecticut, merging with Hartford Carpet and making an unsuccessful bid to form a "company town" just outside of Hartford in Thompsonville. The plan called for New York workers to be invited to join the new company in Connecticut. While some three hundred weavers originally accepted the offer, almost all of them returned to New York within three years. Higgins's search for low taxes and water rates was as well a search for a more docile labor force and higher profits from the fixed capital of Hartford's existing Thompsonville plant.[236]

The relocation of the Higgins Carpet Company, a fixture in Hell's Kitchen and a major employer of skilled male hand-weavers and young, nimble-fingered women, exemplifies the processes of spatial economic forces that buffeted populations dependent on such manufacturing concerns, and is an early example of the geographies of labor that accompany global production systems. Higgins's move and merger were not merely, or even primarily, the result of a search for lower tax and water rates, as the company claimed. The decision to relocate and merge was brought on by Higgins's position within a globalizing economy, affected by changes in the carpet market such as fashion trends, technological advances, and the world commodity prices of necessary raw materials like wool. Higgins Carpet Company's strategic move was the result of multiple factors shaped by a dynamic marketplace in which, as David Harvey demonstrates, capitalist firms faced with falling profits seek a "spatial fix" to solve problems of overaccumulation and labor productivity.[237] Hell's Kitchen residents who worked for such firms were far from isolated populations within self-contained regions, but were in fact global citizens whose daily lives were framed as much by Chinese wool herders as by local conditions. Yet the particular conditions of work, employment, and making a living were hardly a simple process of geographic industrialism. For along with the geographies of labor — the ways that capital uses space to increase profit — there were labor geographies, the multiple ways working people produce laboring spaces to attempt to meet their own needs.

The 1900 relocation of a major Hell's Kitchen employer points out several spatial aspects of the precarious wage-labor market of the period, as well as the nascent world of global capital. The current tendency is to think of such economic relocations as a contemporary phenomenon, a result of the post–Cold War global economy. Jefferson Cowie's work on RCA's search for cheap labor shows that the process of relocation in the face of global market pressures has a longer history, and cities such as Camden, New Jersey, and New York City were part of a larger commodity chain in which the process of linking and de-linking within certain industries greatly affected any city's ability to maintain its tax base and provide municipal services.[238] As David Harvey points out, "the industrial city was an unstable configuration," an outwardly efficient machine for producing, moving, and consuming manufactured products, but also the scene of constant "crisis, caused by overaccumulation, technological change, unemployment, de-skilling, immigration, and factional rivalry." The contradictions of "space-time co-ordinations" of "flows of goods and people" and the diminishing profits caused by disruptions forced urban industrial capitalists to seek what Harvey terms the spatial fix of resource and investment allocation based on maximum profit and circulation of liquid assets.[239] With the rise and fall of small, midsize, and large firms, the local economy of Hell's Kitchen was anything but an efficient system of circulation and accumulation. Like the city itself, the area was more the site of multiple errors and unforeseen consequences contingently connected and disconnected to multiple, often disparate, networks.

The needs of industrial capitalist concerns to maximize profit wrought, as Harvey points out, "far-reaching transformations of all aspects of society." The need to maximize profit for capital accumulation and circulation had profound ramifications for working-class populations such as the residents of Hell's Kitchen, and all major urban areas in the industrial United States. As new types of corporations emerged in the post–Civil War period, their power to influence government decisions increased, as did resistance to corporate domination. As the need for a compliant labor force increased, manufacturing corporations and small businesses recognized the need to reproduce a labor force that was both productive and pliant, at the same time recognizing the importance of alleviating class concerns and preventing potential class alliances. Harvey writes: "It took real political talent and much subtle maneuvering to keep the urban pot from boiling over under the best of circumstances." Thus urban capitalists recognized the need to form new "traditions of community" that could maintain labor control in the face of often dire circumstances, such as the periodic recessions or panics that plagued the capitalist marketplace in the pre–New Deal era. In response, working communities like Hell's Kitchen formed and reformed alliances of identity in the face of the spatial fixes of an often inefficient and unstable market system embedded within a larger spatial framework. While certain processes such as the changing fashion demands of suburban floor covering and the global price of wool were outside the control of the Hell's Kitchen labor pool, other processes, such as organized resistance and alliance formations, were not. As Andrew Herod states, "Workers often made their own space, but not under conditions of their own choosing."[240]

Producing spaces of economic survival in the globalized marketplace of New York City was a daily challenge requiring multiple strategies of resistance, acquiescence, and mobility. However, understanding how space was produced under such conditions does not need to be restricted to merely an examination of the shifting strategies created by market forces. Economic markets and the flows of capital produced serve to construct and frame social space, and encourage or restrict certain strategies. But in studying an aggregate population such as the residents of Hell's Kitchen, concentrating on the vertical, hierarchical scales of production, restricts our understanding of urban space as *only* capitalist space, and confines analysis within the epistemological categories that only allow for a dialectical view of spatial production. While space and social production are embedded within market relations, even the capitalist market is embedded in a physical world that includes the production and reproduction of other spaces. Doreen Massey points out that capitalist industrial space, like all space, is relative and relational, and is best conceptualized "as the product of stretched-out, intersecting and articulating social relations."[241]

WORKING IN THE CITY

It would not be inaccurate to claim that New York was among the first truly industrial global cities, where the urban pot indeed often was in danger of boiling over. The city's strategic location along the eastern seaboard and its deepwater ports made it a natural location as a nodal point of exchange and capital accumulation. In the early Republic, the city competed with other ports, such as Philadelphia, Boston, Charleston, and Newport, Rhode Island, for dominance in merchant trade and finance. The construction of the Erie Canal, linking the Hudson River with the burgeoning interior, and the massive capital accumulation of New York merchants and bankers during the Civil War, solidified New York as the United States' "capital of capital" by the 1870s.[242] The availability of investment capital combined with New York's strategic location and harbor advantages made the city the central location of FIRE firms (finance, insurance, real estate), as well as the home of both small and large industries. Unlike other U.S. cities such as Chicago, Minneapolis, and Detroit, and smaller cities like Rochester and Enfield, Connecticut, no single manufacturing industry dominated the New York economy. Yet the city, by 1890, was home to nearly 30 percent of U.S. industrial production. Though by 1900 the needle trades were by far the largest New York industry, competition and easy access to cheap labor kept most businesses small, with no one manufacturer of ready-made clothing able to dominate the industry.

For the Middle West Side of New York, proximity to the Hudson River meant that much economic activity was centered around the piers and docks, most of which were constructed after 1870. The deep anchorage along most of the West Side provided for not only the docking and loading of steamships, but made the location ideal for portage of rafts of logs from the upper Hudson, and for the Hudson River ice trade. The decision by the city and the newly formed Department of Docks in 1870 to reclaim parts of the riverfront through landfill opened up the area along Twelfth Avenue and West Street to new business. The major gaslight companies that were to merge in 1884 to form the Consolidated Gas Company located their original plants along the waterfront on reclaimed land, employing hundreds of workers to man the lines. The construction of the Hudson Rail Lines depot made the area an ideal location for the slaughterhouse industry, which provided employment, as well as the noxious smells and waste products of that particular industry. The foul nature of slaughterhouse work, located close by residential housing, caused even the moribund New York City Board of Aldermen to act in 1900, limiting the abattoirs to the three city blocks they occupied.[243]

Other small to midsize industrial factories, such as Travers Brothers Twine and Cordage, Nickel and Gross Piano, and Brewster, a maker of carriages, employed between 100 and 400 workers, but none approached the size of Higgins. As late as 1911, according to a factory survey, eighty-two metal works located in the area employed a total of 1,845 workers, an average of twenty-two workers per shop, typical of small-scale production. For women in particular, small manufacturing provided employment opportunities—in candy factories, garment factories, and in industrial laundries. Many men found work as drivers of ice carts and freight wagons, with many making the crossover to mechanized transport with the arrival of the motorized truck. As well, the area's proximity to the theater district east of Eighth Avenue provided jobs for residents as ticket-takers, ushers, cleaners, and even as performers.[244]

Higgins Carpet, employing over 2,000 workers, was atypical of New York industrial concerns, and by far the largest single employer located on Manhattan's Middle West Side. Higgins also differed in its corporate structure and labor practices from the typical Middle West Side concern. By the late 1880s, the company had adopted a governance structure typical of large concerns of the day, with a board of directors making decisions affecting the labor force. It also employed a mix of skilled and unskilled labor, preferring German-American weavers for skilled work, and employing largely untrained young women to manipulate the large looms. By contrast, many employers on the Middle West Side were locally owned small firms, such as local breweries, candy producers, box makers, and garment makers. The labor force of the Middle West Side was mixed, but most workers were unskilled and tended to change jobs rapidly, depending upon local demand and availability of work. The

Spatial Economies 171

PHELAN AND COLLENDER'S WORKS, TENTH AVENUE, NEW YORK.

Manufacturing in Hell's Kitchen at Phelan and Collender. (Picture Collection, The New York Public Library, Astor, Lenox and Tilden Foundations)

dominant feature of work and labor in the area that comes through from many investigations of economic conditions is insecurity and fear of unemployment, both long and short term. Though major firms like Higgins appeared stable and offered the opportunity for long-term employment, smaller concerns such as garment sweatshops and box companies tended to open and close with alarming rapidity, causing insecurity and uncertainty even if a position paying a living wage was temporarily secured. Employment opportunities, particularly along the docks and piers and in the garment industry, were often seasonal, and shifted on the whims of the market. Long-term security was a rarity, requiring workers to adopt and adapt and devise a variety of survival strategies that were often at odds with the demands of capital accumulation.

Strategies of economic survival in Hell's Kitchen were as varied as the backgrounds of its residents. For many, the shifting demands of employers required flexibility, especially within the family unit, as employment insecurity among the male labor force meant that women and children often had to either supplement the male wage or replace it entirely. Ruth True's 1914 investigation of working conditions of the Middle West Side family revealed that nearly 60 percent of mothers were wage earners, either supplementing male wages or being the primary source of income.[245] Of these, more than half worked because the male wage was inadequate to support the basic needs of the family unit. Some of these women "went out" to work, often in sweated or piecemeal labor, but many took in laundry, did sewing, and most commonly served as janitors or basic maintenance workers for the building in which they lived, getting either a wage or reduced rent in return. True documented that most male employment in households was unskilled or semiskilled, predominantly teamsters and dockworkers, whose employment depended on the amount of incoming and outgoing products handled on the docks and piers. Of the fifty-four families True documented, only a small percentage in 1914 could claim to be skilled laborers, such as butchers, machinists, or rail operators.[246] For the unskilled majority, wage rates often did not equal or exceed the basic costs of survival, such as for rent, food, and clothing.[247] In addition to manual labor, Hell's Kitchen was also home to many small businesses. The spatial construction of tenements along the avenues provided room on the bottom floors for businesses that mainly served the local community. Shoe stores, ice providers, butchers, saloons, bakeries, blacksmiths, and other small concerns proliferated. Though many of the owners lived outside the area, a considerable number lived in Hell's Kitchen, often above their businesses.[248]

A result of economic uncertainty was the creation of domestic space as the space of survival strategies. As opposed to the typical middle-class home as the bastion of domestic interiority, the uncertainty of economic conditions in Hell's Kitchen produced the domestic interior and its surroundings as a hybrid of family life and economic production. As in many urban areas populated by unskilled labor, conditions that were thought to disappear with the rise of industrial capitalism, such as the home as workplace, survived long into the new economic order. Many women reported "working out" two or three days a week while supplementing the wage from outwork by taking in piecework sewing, laundry, or small-scale assembly (hat boxes), engaging in child care, performing janitorial work, and the common practice of taking in boarders, who needed to be fed. Of the families in social worker Katherine Anthony's study,[249] 15 percent took in boarders, usually a relative, friend, or coworkers. Washing and ironing clothes, sewing, often on a machine, and janitorial work occupied at least 30 percent of women in Anthony's survey. One

West Side wife and mother supplemented her weekly $5 wage "working out" three days by cooking food for a saloon located on the first floor of her tenement. Whether they worked for a wage or did piecework, few apartments were devoid of some form of labor, from husbands who ran informal shoe repair shops to the infamous home taverns, where for a price neighbors could drink and socialize in the safety of a private residence. As historian Alice Kessler-Harris points out, working-class women "benefited little from the kind of technology that eased the lives of the more affluent."[250] Women "suffered directly from a lack of city services,"[251] having to gather water and fuel (wood and coal) for cooking due to a lack of city water provision and gas lines. Surprisingly, Anthony's study found that most clothing was purchased ready-made, and little home manufacturing of personal items took place.

Many women worked in their tenement buildings not merely as janitors but as unofficial house wardens, compensated with a wage, free rent, or some combination. Some women performed janitorial and superintendent services for several tenements, and preferred this form of labor to outside work. Their duties included cleaning the public areas of the building, collecting ash and garbage, and collecting rent for the landlord. Some women, if they had a good relationship with the landlord, maintained occupancy in the building, seeking out prospective tenants and often screening them as to reliability and ability to pay. Often they would receive bonuses based on this informal labor, and women often thought of a building in which they worked as theirs. They became informal housemothers, overseeing local children, maintaining moral standards, and fighting with neighboring tenants over issues like public cleanliness and consumption of alcohol.

For outside wage work, Hell's Kitchen residents had a variety of options. Male workers predominated in dock work, teamstering, metal work, and wood manufacturing. Female labor was employed in office and house cleaning, assembly production, laundries, garment work, and the production of candy and food products. By the time of Anthony's study, driving trucks and draught teams and dock work were the top occupations for males, laundry work and cleaning predominant among women. For women in particular, daily spatial mobility was a key factor in employment, or outwork. Though some women were willing to travel to jobs cleaning offices in the ever-growing Midtown area, cost and the need to care for families kept many women closer to home when seeking work. As a result, many labored in the local laundries and small manufacturing companies that dotted Hell's Kitchen. As Kessler-Harris notes, the decision to perform outwork was for many women of the period based upon "a changing tapestry of expanding opportunity and shifting household needs"; available opportunity and often necessity greatly influenced working women's "self-perception and aspirations."[252]

For those willing to travel, cleaning offices and theaters between Fourth and Seventh Avenues from Twenty-third Street north into the Fifties was an employment option. Approximately fifty theaters and concert halls stood within walking distance of Hell's Kitchen, and new office towers proliferated. In addition, schools, hospitals, churches, private clubs, and residences also provided cleaning jobs. While the work was difficult and often damaging to the body, it offered in many cases more security than other types of employment, as hospitals were less likely to close and move than small factories, and large office buildings rarely went unused in light of the investment necessary for their construction. However, much theater work, such as cleaning, ticket taking, or ushering, was seasonal, dependent upon New York's theater schedule, since performing companies traveled in summer months. One enterprising woman in Anthony's study, single and without home responsibilities, countered this trend by taking summer work cleaning in Atlantic City in New Jersey.[253]

Public cleaning was a difficult job that exacted a toll on women workers. One testified that she went home and cried after work because "her knees hurt her so."[254] Applying Vaseline, on a friend's recommendation, only made her knees "softer" and increased the discomfort. A recurring complaint among cleaners was the inability to function upon returning home due to exhaustion. One woman reports that she missed her second child's first steps due to such conditions. When she was home, working the irregular hours of many cleaning jobs, her child, cared for by neighbors and her sister, was usually asleep. Yet despite the drawbacks, taking public cleaning work was often a better option for cleaning women than the older model of domestic servant, whose duties often included "living in" in private homes. Although the pay for inside cleaning labor was often higher, and meals and gleaning of surplus food were usually perks of the job, most Middle West Side women preferred working out at cleaning for a variety of reasons. Live-in work had also become increasingly scarce, due to the increasing proliferation of "French flats," multi-storied apartments for the upper-middle classes, which contained no bedrooms for servants. Most women preferred or needed to return home after work, and many did not appreciate the constant attention and supervision that went with live-in work. As well, the physical environment in which women performed public cleaning was relatively free of the noise and extremes of heat encountered by women doing factory or laundry work.

Service work in hotels and stores and clerical work in small businesses were other work options for women that presented their own spatial options. Hell's Kitchen women who chose work in area hotels, bars, and restaurants faced difficult choices that reflected community mores. For these women, respectability was often questioned based on the nature of their work and its proximity to questionable social behavior. Hotels of the period often served several purposes, as gambling houses and spaces of illicit assignation, and women who worked at such concerns had their morals questioned by others in the neighborhood. Anthony reports that many women, especially those who were married, lied about their employment to neighbors and associates to avoid the stigma of being known to work in an immoral environment.[255]

Closer to home, laundry work was a primary occupation for West Side women. Privately owned laundries, using machines that needed tending, dotted the Middle West Side. Women reported that though ownership of the laundries shifted rapidly, the work itself tended to be steady and relatively immune to market fluctuations. The heat and noise from the machines created a physically demanding environment that contributed to high turnover, making the laundries a prime site for women seeking temporary forms of wage work, such as when a husband was injured or laid off. Due to the constant demand for labor, Middle West Side women viewed working in industrial laundries as either long-term steady employment, though one that took a physical toll, or as temporary labor readily available. In Anthony's study, Mrs. Myers, married with school-aged children, reported working in the same laundry through three ownership changes, as her relative expertise and experience with ironing shirtwaists, and relatively low wage demands, made her a bargain for each new ownership group.[256] Conversely, another survey participant reported working for sporadic periods in the local laundries, particularly after her marriage, when she picked up work to help out in emergencies.[257] Hell's Kitchen women reported that acquiring laundry work was often dependent upon being aggressive when seeking a job. It was the "pushing sort" who more easily found positions, and industrious yet quiet, unassuming women were often at a "distinct disadvantage." Mrs. Snyder, described by Anthony as a young woman with a "hustling manner and engaging smile," declared that she could "always get taken on" due to her outward display of "ambition."[258]

Ambition and self-perception were an integral factor in the spatial world of the laundry and other labor sites. The labor process in laundries was divided into sites dependent upon skill, strenuousness of work, and wage. The different labor processes included mangling (feeding the machine), shaking (after washing), hand ironing, starching, collar-machine operating, marking and sorting, and supervision. Mangling and shaking were the positions requiring the least skill and training, and were the jobs most likely to go to women seeking short-term employment. Mangling, the process of feeding clothes into a machine, was considered the least demanding "on the muscles," but was noted for its monotony and location near the heat and moisture of the machines. It was a job held mainly by younger women, who were most apt to move on due to boredom, lack of necessity for steady work, too many demands for overtime,[259] or finding a better job.[260] Shakers did what was considered the most arduous and physically taxing unskilled job in the laundry. Shaking required repetitive shaking out of wet sheets, heavy linen, towels, coats, and aprons, and involved "heavy lifting, constant stooping, bending and reaching." It was also the lowest-paid position,[261] and was often taken up by "older women who were forced to go out and earn" due to a temporary home condition.[262] Anthony reports the case of one woman, named Harrison, who was offered training by the Charity Organization Society in the higher skilled and better-paying position of ironing, but who gave up the trial and returned to shaking because she deemed herself "too old to learn." Conversely, women considered to have ambition, often defined as some combination of a willingness to ignore the physical toll of labor and a desire for higher wages based on faithfulness and industry, worked in the more skilled positions such as ironing and collar work. These women, who sometimes could move up to supervisory positions, were by and large "proud of their stable work records," often in spite of an inability to be mobile within the industry. Many took some pride in their ability to produce fine finished products such as well-starched and ironed shirts, and looked down on younger women who frivolously changed positions.

The female labor force of the Middle West Side, working within the lived space of daily experience, space not of their own choosing, exercised a certain degree of choice and autonomy in their labor, and often produced spaces of sociability in their daily rounds. Women's labor was, of course, crucial to the household economy, and though many would have preferred the role of housewife that was a full-time job in itself, those who performed outwork often took pride in their labor, judged others by the standards of hard work and dependability, and gained a measure of autonomy and experience otherwise unavailable. Their spaces of labor were not their own creations, but instead dependent upon a wider marketplace of profit and production. Industrial laundries were created to service middle-class apartment-dwelling New Yorkers whose homes contained no facilities for such services. It was also the result of the values of these customers, who considered it improper to have laundry done in the home. Laundries also serviced institutions, such as hospitals and the New York Sanitation Department, whose crisp white uniforms symbolized their mission. Theaters served a similar middle-class audience, whose tastes and values determined the success of a given performance, and thus the length of a show's run, often affecting work opportunities. Candy and garment factories were likewise buffeted by shifting middle-class tastes, as well as by the availability and price of global commodities that made up the raw materials of the production process, such as sugar, cocoa, and wool. While the meta-space of the marketplace was beyond the control of working women, and the shape, design, and layout of outwork locations were determined by other factors, the spaces of labor took on a variety of meanings for Hell's Kitchen women. The meaning assigned and contested in labor space has much to say about women's roles in the Hell's Kitchen economy, their relations to their own bodies and their own narratives, and the shifting assemblages of alliance and fragmentation that constituted Hell's Kitchen's spatially constitutive community.

Katherine Anthony's 1914 study of Middle West Side women's labor begins its chapter "Occupations of the Mothers" with a bleak description of the presumed indignity of public office cleaning: "The appearance of the charwomen on their knees scrubbing an office floor, a public corridor, or lobby of a theater is not one which inspires respect in the ordinary passerby."[263] Aside from the "ancient stigma" attached to cleaning work, and because of the public nature of the labor, a "familiar sight" of the business district, office cleaning was deprived of "any measure of human dignity." Anthony clearly expresses a class-based gaze on the space of the worker, further romanticizing the imagined scene by reminding her readers that a "human life" is "imprisoned within each of these humble figures," women who fight a daily battle that "takes more courage than Waterloo." Anthony's description paints the spatial imaginary of the heroic working-class woman, backbone of the struggling family, anonymous in the public sphere, yet wizened by her years of toil under the most inglorious of conditions, the very image of the deserving poor.

Capturing Anthony's gaze, combined with the considerable physical evidence regarding women's labor during this period, it is difficult to dispel some of the truth of her descriptive effort. Public cleaning of offices, theaters, hospitals, and private homes was difficult work. In descriptions like Anthony's, and in the women's own testimony, political scientist Iris Marion Young's "five faces of oppression"— exploitation, marginalization, powerlessness, cultural imperialism, and violence— are clearly present.[264] A large available labor force kept wages low and hours long, and this factor and the dispersed nature of the work mitigated against organizing, encouraging exploitation and lack of power over wages and conditions. Anthony reports that many office workers considered it a kindness to "neglect and ignore" the women in their tasks, the very definition of marginalization. Employers' emphasis on punctuality, appearance, and deportment for a labor force that often had not the time or resources to easily conform provided a measure of cultural imperialism. And the toll on the body, in the form of bone, joint, and ligament damage, added to breathing in fluids, dust, and germs, adds violence, as well as what Harvey describes as ecological consequences of oppressive employment conditions.[265]

Yet the image of the charwomen scrubbing ceaselessly in exploited, marginalized indignity is not the sum total of the work and spatial experience of Hell's Kitchen's laboring women. Not discounting the difficult nature of the work, women in Anthony's study also reported various levels of motivation and agency in their choice of employment. One Middle West Side woman reports her choice of employment to be based on her relative susceptibility to different climatic conditions. She had given up laundry work because of the noise and heat of the machines for employment in a chocolate factory, where her duties involved placing nuts on chocolate as it congealed in cold temperature. She had decided she could tolerate the cold better than the noise, heat, and moisture.[266] For many Middle West Side working women, decisions on what type of employment to engage in came down to such basic choices. Exercising agency in employment was often an issue of the real or imagined effects on the body, and the horizon of opportunity that each position held. For some women who were content with temporary labor to fill an income gap, unskilled work in a laundry, where the bodily toll was high but the expectations were low, was the best choice. For others, whose self-conception was dependent on their position within the labor force, skilled long-term employment matched their personal conception of ambition and their daily values. We can also imagine that differing spatial experiences held multiple meanings for these women, as forms of labor segmented their daily lives, forming potential conceptual spaces of difference.

Primarily spaces of labor-segmented women's lives moved between the traditional sphere of the private home and the public space of employment. Working women also experienced space in the workplace as segmented and produced, as the example of the laundry demonstrates. The production of space in women's labor in Hell's Kitchen can be analyzed with various scales. Whereas economic conditions such as the rising demand for office space and the removal of laundry work from middle-class domestic space were key factors in the production of laboring spaces, feminist geography has examined such produced labor spaces through different scalar approaches. The work of Isabel Dyck and other feminist geographers intensifies the epistemological gaze by moving from the macro-structural scale to the "taken-for-granted mundane, routine activities" of the everyday, which she terms the "hidden spaces of place-making."[267] The question arises, though, from whom were these spaces hidden? Certainly not from the participants, who produced the laboring work spaces under conditions not always of their own choosing. To accept this argument, that women in mundane jobs create hidden places of performance, we must examine our notions of place and how the term is understood, as well as received and reified notions of economic scale.

Dyck and other labor geographers utilize notions of vertical scale to describe and analyze the production of place making in the work process. In this mode of analysis, larger then decreasingly smaller economic scales descend in an interdependent vertical axis to produce the local places of identity, where working people create the places that in some ways are compensatory for the injustices, exploitation, and accumulated indignities of the vertically integrated work process, or where they can develop and articulate modes of resistance based on a common exploitation. In this analysis, working space is reclaimed as place, as workers exercise their own agency to redeploy working space despite the uneven power relationships that govern the vertical scales producing the work environment. Though there is little doubt that work conditions were physically and mentally challenging, and various forms of exploitation were produced by the scales of economic production, examining working spaces as contingent, horizontally integrated sites can offer a different view of the production of work space, and a different way to describe and analyze workers under conditions not of their own choosing. A brief look at two worksites and their effect on the laborer, the male-dominated waterfront docks and the female-dominated site of textile labor, illustrates this point.

THE LABORING BODY

IN HIS EXHAUSTIVE STUDY of dock work carried out in 1911, Charles Barnes states that the unregulated conditions existing along the Middle West Side piers were directly responsible for the "distress and dislocation of healthy community life" that so aroused the reformscape gaze.[268] Frustrated former Commissioner of Docks Calvin Tompkins expressed similar sentiments in 1905, decrying the city's lack of control over one of its most prized and valuable assets: "At smaller ports the dependence of city growth upon terminal facilities is beginning to be appreciated and acted upon. But in New York, the largest seaport of the world, there is no such popular understanding of this fact, nor the city's needs for schools, sanitation, water supply, pavements, police."[269] Tompkins states that the city will never fully realize its economic and social potential until it is freed from the "pressure of corporate interests."[270]

Dock work, the labor of the longshoremen, was male labor, strictly segregated by gender. Work on New York City's docks, like work in public cleaning, laundries, and small manufacturing, was conditioned by scales of economic and political activity, and male labor space was produced within this nexus of factors. Male labor space on the docks and piers was structured by the natural and built environment, and conditioned by such disparate factors as federally regulated tariffs on imports, city administration of infrastructure, and public perception of longshore workers. In addition, like spatially segregated women's labor sites, male-dominated pier labor produced not the hidden places of compensation, but horizontally layered locations that both solidified and fragmented the laboring force.

The economic and political factors that contributed to the production of the chaotic and degraded spaces on Hell's Kitchen's docks and piers would require a separate and much larger study to enumerate.[271] The critique of the lack of city control given by both Barnes and Tompkins reflects the lack of political and economic oversight governing this vital infrastructure during the period. The city created the Department of Docks in 1871 in a late and largely unsuccessful attempt to regain control of the commercial waterfront dominated by private ownership.[272] In theory, the legislation creating the department returned waterfront control to the city administration, but entrenched private interests continued to rule, consistently challenging the city's authority, and often prevailing, particularly in legal cases of proposed condemnation of structurally insufficient facilities.[273] As a result, by 1900 the city directly controlled only five piers in the Middle West Side, while many of the privately operated facilities, as Tompkins laments, were both inadequate to handling an expansion of trade, and inefficient and dangerous.[274] For example, the city and Department of Docks engaged in ongoing struggles with private owners over the conditions of their facilities, at the same time suffering from constant budgetary constraints in attempting to improve conditions. In 1896, the city waged battle with two Middle West Side companies, Knickerbocker Ice and Metropolitan Manure, over conditions on their piers. Knickerbocker Ice had allowed dangerous conditions, such as erosion to the underlying structure, to persist, and Metropolitan Manure engaged in dumping of offal and other waste directly into the waters off the Middle West Side.[275] In both cases legal rights of the property owners left the city powerless to improve conditions. Tammany Hall's own George Plunkitt,[276] the owner of the 51st Street pier, was unsuccessfully sued by the city that year for failing to make needed improvements. Repeated requests by the Department of Health to dredge the Middle West Side piers before the summer swimming season went unanswered due to lack of budget. Nearly two decades later, in 1913, Calvin Tompkins reports that conditions had not improved. No "organic" plan existed to speed the efficient movement of goods and ships, and dredging and waste removal proceeded at a glacial pace. The 1912 Department of Docks report clearly states the problem at hand when discussing the need for expanded facilities to handle passenger ships. "How far can the Department rely upon the power to condemn private property?"[277] The impact of ineffective regulatory oversight affected the physical conditions of the infrastructure as well as the health and well-being of those who lived around it. As Calvin Tompkins pointed out, "bad industrial conditions imply bad living conditions" for the waterfront workforce. The unregulated nature of the docks and piers contributed to outside perceptions of Hell's Kitchen, and further affected the self-conceptions of local residents.

The Middle West Side waterfront was also subject to changing conditions in the marketplace, and dependent upon the amount of import and export of goods moving through its facilities. Also, many of the piers in the area were linked to localized commerce, such as waste disposal and ice distribution. For the immediate area, the 1907 Department of Docks report lists twelve working piers between 37th and 56th Street, with six utilized for local or city economic interests, such as ice, manure, a recreation pier,[278] a city-owned ferry terminal for New Jersey traffic, and a city-owned tool and utility shed for dock maintenance and repair. Two piers were used for passenger lines, and three were dedicated to long-distance freight, including the 40th Street piers of the Union Stock Yard and Market Company.[279] Of course, not all Hell's Kitchen workers employed on the docks worked exclusively in the immediate area; much traditional longshore work was located on the downriver piers of Chelsea, the East River, and Brooklyn. Expansion of the city's transport infrastructure in the first decade of the twentieth century increased opportunities for work outside of walking distance, but a good number of Hell's Kitchen residents did find employment on the neighborhood docks. In 1907, of the 2,000 employees working on docks and piers in the city, some 230 can be identified as from the Middle West Side. Indicative of the economic conditions of the area, of these a mere seven were employed in what could be considered white-collar or managerial positions. While some residents worked in skilled positions on the loading docks, many more worked in unskilled jobs in general maintenance and support of the loading crews. Of Hell's Kitchen residents, the highest-paid dockworker for 1907 was John Wood, an engineer who earned $1,800 for the year, a figure that would place him barely above the survival line of the standard of living studies of the day.

The general conditions of the pier areas and the sporadic and inconsistent nature of waterfront labor contributed to the perception of the piers as a place the respectable avoided, contributing to the overall reputation of the Middle West Side. While less skilled pier work such as hauling and dumping trash tended to be more consistent, lower-paying skilled labor in loading and unloading was inconsistent, dependent upon the import-export levels of the global commodity markets. As a result, daily employment for dock workers was determined by the "shape-up," in which experienced workers gather to be called to work by crew organizers or stevedores, based on the arrival of a laden ship or the need to load a departing vessel. For Barnes's study, the shape-up is the primary space of dock work. As the telegraph message[280] announcing the imminent arrival of a ship is received, a flag of arrival is raised on the pier, and word spreads among potential workers that "the flag is up," and a crew needed. According to Barnes's report, the call to shape up brings the men together in the open space in front of the working pier, space inadequate to hold the workers in the traditional semicircle of shape lines. The men jostle for position, joke and occasionally scuffle, and await the decision on who will be called to the crew. Stevedore foremen call men based on experience, reputation, or on recommendation. In certain circumstances, workingmen pay kickbacks to hiring agents who guarantee them a working day. At some piers, where the work is specialized, shape crews are divided by duties, with separate groups for hold men, deck men, and pier men. These are duties that require some skill and training, and are often chosen by "gang"—men who have worked together in the past, and know each other's skill level and specialty. Dock work, like laundry work, is divided by skill and pay level, and different jobs are seen with different levels of status and respect.

The hiring space described by Barnes as the start of the work process is a highly complex and almost baroque production of space. It is also the culmination of the waiting interval, the time between paid jobs that contribute to the culture of labor, masculinity, and hierarchy typical of male-dominated employment of the period. Intervals between work and the nature of the shape-up also contribute to the outside perceptions of dockworkers, something Barnes comments on at length in his study. Because dock work relies on incoming and outgoing ships that arrive and depart, not on the workers' timetable, dock workers often spend long hours, or even days, waiting for work. As a result, men often have little to do but seek other menial employment in intervals, or spend time near the docks with other workers. These physical spaces are layered with meaning, for both insiders and outside observers. For the workingmen, shape-up space is where hierarchies are contested and negotiated, often through rough joking and play. Ethnic, racial, and skill divisions are prevalent, and often resentments and insults can lead to physical violence. To the outsider, the mass of men shaping up, though divided by highly contested and established hierarchies, can appear as a homogenous collection of unskilled and unsophisticated brutes, mirroring many outside perceptions of districts like Hell's Kitchen. This perception is often reinforced in the forced periods of inactivity, when men often congregate at or near saloons, or in public space, appearing rough-hewn and rowdy, the improper citizen *par excellence*. Barnes lays the responsibility for the lack of attention paid to this important labor force and their working infrastructure directly to these widespread public misperceptions of the true nature of dock work. "By people generally, the longshoreman is classed loosely with the industrial outcast, the unskilled laborer. He is thought to be shiftless and a drunkard, as unworthy of serious attention."[281] Thus, though the space of New York's docks and piers are in plain sight, Barnes reconstructs them in his study as hidden places, obscured by the inability of the reforming and political classes to seriously consider the plight of the workingman. They become analogous to the hidden places of Dyck's feminist geography, where the noble but exploited workers create spaces for the performance of layered identities that somehow compensate for the difficult conditions of survival.

The space of dock and pier work, and its proximity to other spaces, contributed to this perception. The many colorful characters attracted to waterfront industrial sites also framed public perception. The polluted waters of the Hudson River and the stench of manure, offal, and other olfactory delights emanating from the 40th Street slaughterhouse gave the area a foul smell in general. As Tompkins points out in numerous reports, the physical conditions of the pier structures, often in disrepair, contribute to the foreboding sense of the environment. Dock work, though specialized at the level of loading and unloading holds, also requires the hiring of casual labor, unskilled workers hired at very low wages who handle basic cleanup and menial tasks. Called "shenangoes" by Barnes's longshoremen, these workers are often drawn from those too old or sickened by disease and alcohol to work regular crews, or from men who hang around the docks in search of sexual encounters with ship crewmen and dockworkers. The physical deterioration of the piers, and the composition of the crowds of men congregating, made most docks a no-go area for the respectable public.

Barnes's 1913 study, sponsored by the Russell Sage Foundation, attempts to bring visibility to this occluded situation. In the preface to his work, Sage director John M. Glenn expresses the hope that more attention to the uneven and exploitive nature of dock work will inspire the public to "get into the game" and "through the weight of public opinion and legislation compel steady improvement" in waterfront working conditions. What Barnes points to in his study is the craft, skill, and danger of the work as well as its importance to the larger economy.[282] Though his study is intended to "shed the light of reform" on the neglected conditions of the working piers, what it ultimately reveals as a historical archive are the multiple and overlapping boundaries of horizontal spatial production.

Barnes is more prescient than he perhaps knows when describing the conditions of the working piers and the men: "To what race, to what nation, does the longshoreman of the port belong?" He goes on to describe the "cosmopolitan character" of the workforce, the "quaint jargon" that is "almost a dialect" in its "suggestiveness and variety."[283] The description accurately portrays the multiple sites of spatial production existing among the waterfront workforce. Though ethnicity and race, job status and pay, and the relative skill and care brought to the different work processes were certainly boundaries of demarcation and identity, the multitude of factors that determined these categories, as well as the variety of ways masculinity was performed, made the boundaries porous and crossable among a diverse workforce that developed a jargon of communication necessary to work and survive under dangerous and uncertain conditions. Workingmen arrived at the piers from a variety of backgrounds and experiences, with different levels of skill, pride, knowledge, and with different goals concerning the working day and the laboring experience, and spent their working and down time not in strictly demarcated places, but in differently situated sites of survival and sociality.

The primary site of dock work was the shape-up, the gathering point for potential crewmen who worked on incoming and outgoing freight, and Barnes describes it as simultaneously the site of conviviality, tension, ethnic, racial, and skill segregation, and a space for the performance of masculinity, which in the case of dock work, manifested in displays of strength, competence, mastery, leadership, and the ability to defend oneself in disputes. Workers gathered near friends and ethnic associates, but just as often men would gravitate toward other men who would surely be chosen for any crew, either through reputation or previous arrangement with the stevedore. Workers whose numbers were called early in the shape-up often petitioned the stevedores to accept their associates into the work crew, based on their recommendation. Although these associations were sometimes based on ethnic solidarity, they were just as often based on skill and reliability, as the experienced men would want others on their crew who would make their own work time more enjoyable and less dangerous. Ethnic solidarity was important to some crew members, but the dangerous nature of cargo handling made experience, sobriety, and safety qualities that could potentially override these solidarities, leading to tensions and rivalries among members of ethnic groups.[284]

Once the crews were selected, the work process and division of labor took precedence over other considerations. With the demand for speed and efficiency, loading and unloading cargo required a sophisticated division of labor, exhaustive stints of labor with brief respites, under all manner of weather conditions, in an environment of fast-moving, extremely dangerous cargo. The most frequently mentioned condition, by both Barnes and his subjects, was physical danger, the hazards of working with fast-moving parts under conditions of stress.[285] The proximity to swinging cargo loads, slippery stairs and gangways, sharp edges, and poorly lit holds, combined with the relative experience or inexperience of crew members, frames the longshoreman as a member of a working organism.

Reading Barnes's summation of the spatial conditions and the workers' accounts gives the reader a feel for the accelerated temporality produced at this site of emergence, as networks overlap in a repeating yet contingent composition of work process and work space, in which a wrong motion or mistaken movement can result in injury or worse. The dangerous conditions, the demand for speed, and the need for the crew to act as a unit, both in terms of the demands of their employers and in their own interests, produced work spaces that were the result of multiple, horizontally overlapping networks of composition. For example, in the process of loading freight, included are the processes of shipbuilding the collection, packaging, and shipping of the cargo; the origin of the cargo (commodity crops, industrial products, building materials, etc); the navigation of the waterways; the waterways themselves; the aggregate labor force; their clothing, to take just a partial account. As the hatch gang, the group of pier, deck, and hold men, proceed with their work, the relative temporalities compose the temporary work site. The loading process requires skill, precision, timing, oversight, muscle, and a certain confidence in fellow workers. Barnes's description of a load lowering into the hold provides an idea of the temporality:

> While the first drumend man has been raising the draft, another is taking turns around the second drumend with his fall, which has been attached to the first fall, preparatory to swinging or lowering the load. When the signal for lowering is given, the speed of the winch is lessened and the second drumend man slackens his rope and lowers the draft into place in the hold. When a draft is being

lowered the slack of the rope will pass rapidly through the hands. To protect himself, the drumend men wear leather hand guards like a half mitten, loosely fastened at the wrist, which he swings in front of his palm as needed. By the time the load has cleared the rail and arrived over the hatch, the gangway man is there ready to signal, first for the men in the hold to stand clear, then for the lowering of the goods. Sometimes, the drumend man has a telltale mark on the fall to indicate how far he should let it run; but more often knowing on which deck in the hold his draft must land, he tells by feel how far to let it go.

At the signal of the gangway men the draft is let go with a rush and is stopped suddenly a few feet above the deck on which it is to be stowed. It is then seized by the hold men and swung over as far as possible toward the position the goods will finally occupy. The gangway man, by watching or by means of a signal from the hold, learns then it is "ready to let go." Again he signals the drumend man, and the draft is landed. The sling is then thrown off by the hold men, and the empty sling is fastened to the hook of the fall as it rises from the hold to be lowered to the pier. Meanwhile, the hold men with hands, hooks or hand trucks are hauling and rolling the goods and fitting them into place.[286]

The process is repeated until the hold is full. If not done properly, injuries ranging from "a crushed toe to a fractured skull" can result, including numerous cases of permanent maiming and death. Damage to the shipment can result in lawsuits and in members of the crew being blackballed by stevedores. In worst-case scenarios for the shippers, an improperly loaded cargo can result in a capsized ship or pierced hull, but such cases were rare. By far the most frequent negative consequence in the process was personal injury to members of the crew, resulting in loss of wages, and sometimes permanent disability. The multifactored work site produces spaces for different levels of performance. Skill, masculinity, ethnic solidarity, and tension, trust, apprehension, and exhaustion are only a partial list of the differing affectual forces within this work process. In two examples, the shifting boundaries of solidarity and fragmentation are apparent, produced by the laboring space. First, the ethnic camaraderie played out in downtime is likely to disappear if a member of one's identity group fails to perform his assigned task and causes

Spatial Economies 189

Stevedores working the docks of the Hudson River on the Middle West Side. (http://www.archives.gov/)

injury to another crew member. Conversely, ethnic lines dissolve if a crew member is seen as a skillful and dependable performer. As one form of solidarity is fragmented, another type of fragmentation is overcome by the spatial needs of a dangerous work process.

Second, the performance of masculinity on differing levels plays out among the physical objects and accelerated temporality of the process. Although most of the men Barnes interviewed agreed that their work environment was dangerous, their reactions to the work space displayed differing relationships to the hazards of the work. As Barnes states, the "longshoremen differ widely in their estimate of the risk of their work, and the length of a man's working life." For some, reporting injuries that were not directly incapacitating slowed down the work process and shortened the valued downtime. Many dock workers failed to report minor accidents for fear of being mocked by fellow crew members or blamed for a longer work shift. Others reported every minor bruise, and many were consistently involved in court proceedings against ship lines to attempt to gain compensation for injuries. Such men were considered less than manly by many of their coworkers, yet role models and leaders by others. Typical was one man who claimed to Barnes that he had never been hurt on the job, until his daughter brought up his "severed toe."[287] Longshore workers generally exhibited a certain pride in being able to withstand and even prosper under the given conditions of their work process and often produced spaces of accomplishment and arrogance, though the temporary nature of this space was decomposed when the work process ended. In sum, multiple masculinities, not a stable masculinity, were performed in the accelerated tempo of the work processes, and entry into a spatial system where gender identification is merely one of many modes of being available, some chosen and some more random, is based on the worker's position within the process.

Downtime, the time between jobs, decelerated the temporal pace of the long work stretches, which could last up to thirty hours of continuous labor, and was produced under different network conditions. Downtime, when dock workers would often congregate near the piers anticipating the arrival of new ships, contributed to the reputation of the waterfront, and was produced by networks of market connections, political alliances, and the natural conditions of the river. As Barnes states, little spatial provision was made for longshore workers on downtime. He reports that conveniences such as clean toilets, washrooms, resting areas with couches or chairs, were "almost totally lacking," causing many men to "hang out" on streets or "sit on freight piled in front of the piers." He describes the men on break as "gregarious," and sees "skylarking," the practices of pitching pennies, reading newspapers aloud, or engaging in "friendly" wrestling matches as a "safety valve" against the extreme pace of the work period. "Once all the pressure of work is removed, they are like boys on a holiday." Several failed attempts had been made to establish shelters for men on downtime, and had failed due to lack of funding and oversight. In 1910, the Church Temperance Society finally established a working shelter for Hudson River work crews at West 22nd Street, but at the time of Barnes's report, was still "viewed with suspicion" by many of the men, and had been vigorously opposed by the owners of the small taverns that lined the pier area from lower Manhattan to the Middle West Side. Although many of the men refused to utilize the Church Temperance Society space, others took full advantage, using it to rest, socialize, or read in the free library.[288]

The spatiotemporality of downtime produced solidarities and fragmentations within multiple moving networks, sometimes coagulating around a shared interest, such as the establishment of an organized and subsidized resting space along the piers, or fragmenting over personal disputes produced by the long waiting periods, frustration, exhaustion, and proximity to alcoholic beverages. But even though many of the men preferred and bargained for rest spaces where they could await the next work shift, others opposed such spaces as disciplining and constricting, and where rules regarding the consumption of alcohol normally prevailed. Though downtime could produce space whose much slower, even monotonous, temporality was conducive to camaraderie, it could just as easily produce spaces of conflict and dispute. Barnes reports that card games often turned violent, and that workers who entered the rest space already intoxicated, though not required to leave, were "discouraged" from staying, often resulting in physical altercations.

Unloading cargo ships was only one part of the process of spatial production in waterfront labor. Shape-up work was sporadic and created its own perception among the workers themselves and outside observers. Other forms of waterfront labor often stood in contrast to longshore work, despite several similarities. Though many who labored on the waterfront experienced conditions similar to those of the longshore crews, their conditions of employment, experience, background, and the relative stability or instability of their positions produced different spaces. Other forms of employment along the waterfront ranged from semiskilled drivers of horse carts (and later, motor trucks) to dock supervisors, who were considered white-collar labor.

Labor solidarity and fragmentation also produced space on the waterfront. Dock workers had a long history of participation in the movements of organized labor, with varying degrees of success, failure, solidarity, and mistrust. Classic labor history viewed questions of solidarity through the relative levels of institutional organization and what historians deemed radical labor action, such as strikes and other forms of resistance.[289] For the New York waterfront, this type of organization was uneven, rising and falling with market conditions. The produced space of the labor process played a key role in organized action, as the experienced crewmen, temporally loyal to their union, believed that crews of ordinary men could not perform the labor necessary to the loading and unloading process. In fact, the strike of 1907 was held off until July of that year, with the workers believing that just the smell of the hold in midsummer would drive off temporary replacements. Two results of the space produced by organized labor struggles on the waterfront were a general distrust among longshore workers of union officers, and resentment of local police for siding with management in disputes.[290]

Uncertainty was the dominant condition produced by waterfront labor space, as it also was in the mixed-gender spaces of factory labor, in the female-dominated labor spaces, in the home, and in the interstitial spaces, such as the shape-up areas and break rooms, of the Middle West Side. Numerous social survey reports document the various ways uncertainty of employment, the constant threat of dispossession, had on Middle West Side workers. Uncertainty could appear in the form of the closing of a local business, the seasonal cycle of certain forms of employment, injury, age, changing technologies in the labor process, and changes in the domestic situation. The networks of social contacts and strategies of survival, and sometimes of security and relative prosperity, operated within the framework of other networks, variously linked and decoupled from the local. Yet any study that attempts to accurately describe spaces of labor under such conditions is limited if it is restricted to understanding only the economic market connectivity and the mediated forms of social reproduction emanating from such connections. Such a view also assumes a degree of connectivity that structures the identity of human actors and their "interests" in both the labor process and the production of space. A return to the story of Higgins Carpet illustrates the process.

Higgins Carpet was atypical of Hell's Kitchen industries when it closed its doors in 1900 and moved its operations to Connecticut, abandoning the Middle West Side in search of literally greener pastures. In one sense, Higgins's decision to relocate can be viewed within the framework of the multiple-scale economy operating at the turn of the century, in which global commodity prices, local labor costs, and other factors working at interconnected scales affect the conditions of possibility, or life chances, of the working population. Many factors figured into the Higgins decision, including its shift from single-family ownership to corporate board governance, the changing market for floor covering, and the search for a pliant labor force.[291] All of these interconnected factors were the product of a produced spatiality, the timing and geography of capitalist business practice. All were as well network assemblages that included non-human material factors producing in their interaction with the human factors the sites of emergence containing indeterminate potentials for temporary outcomes. For example, the accelerated tempo of cargo work could be severely interrupted by faulty equipment or worn rigging, or halted completely by inclement weather or poor navigational conditions that delayed the arrival of ships. So for a given sector such as the carpet industry, global market forces, technological changes, labor market pressures, and the raw materials used in the production process combine to create durable if always changing conditions. A brief description of a labor dispute illustrates these points.

Just one year before Higgins Carpet made the decision to vacate Hell's Kitchen, management and labor engaged in a dispute over working conditions produced under the spatial conditions of market forces and industrial technology. Alterations in the large looms used to produce carpet fibers, many already instituted by Higgins's smaller competitors, were introduced on the factory floor in late 1898. The technological improvements in the machine process allowed one worker to operate two machines simultaneously, increasing productivity while increasing the pace of the factory floor and changing the movements of the machine operators. Higgins, like their large competitors Bigelow Carpet and Sanford Carpet, had long resisted the installation of new machinery, having made large capital investments in the 1870s, only to watch themselves and their competitors go into a downturn as the market was flooded with carpet.[292] As smaller competitors, particularly in the market for mass-produced carpet, increasingly cut into Higgins's profit, the company installed in-grain machines in 1890s, hiring largely unskilled women to tend them. The women labored in the same building as skilled weavers who produced hand-woven luxury carpets, and who were considered more radical on labor issues, as well as being better compensated. But the installation of new machinery in 1898 produced the accelerated pace of the work floor that turned the unskilled factory girls into radicals.

The strike of 1899 against Higgins was carried out by over two hundred unorganized, unskilled workers over changing conditions on the factory floor. Approximately 1,600 other employees, unaffected in the immediate sense by the new machinery, refused to join the walkout, though most were sympathetic to the striking women. The walkout started on January 20, 1899, and was settled by the end of the month, when the Board of Arbitration and Mediation of the State of New York brokered a deal between the opposing sides.[293] The conditions that brought on the labor action and eventual arbitration were produced by multiple causal factors involving both human (wages, work conditions) and non-human agents (new machines), operating in relatively autonomous networks that overlapped to compose the site of the event. The move by Higgins to install the new machinery and increase the workload of the tenders was accompanied by a corresponding increase in the piecework wage by 15 percent, intended as an inducement to the workers to accept the accelerated pace of their labor. Each worker, typically a young woman, was now required to move between two machines rather than be stationed at one. One additional worker would be added for each ten machines to cover breaks and correct errors that resulted from the increased pace. The installation of the new work process altered an already loud, fast-moving, and often chaotic shop floor, adding new movements, bringing more raw material to the floor, and increasing the pace and amount of material, finished product, and working bodies.[294]

The work crew quickly rejected the new process despite the increased wage it promised. The girls claimed that any increase in the amount they could produce, thus increasing their wage, would be offset by the increase in downtime made necessary by the new process. The increased pace of work was, in any case, not considered a fair trade-off in the view of the majority of the loom tenders. In the arbitration hearing, two workers representing the strikers, Rose Cain and Lizzie Irving, testified that the increased pace made it "physically impossible" to gain the increased wage without "breaking down." Irving further testified that Higgins had resumed using powder dyes that had been forbidden by the Board of Health. It is not clear from the proceedings whether or not she volunteered the information regarding the dangerous dyes, or if she would have reported their use without the forum provided by the mediation of the strike.[295] What is clear is that Cain, Irving, and the other two hundred strikers objected to the alteration of the work pace created

Working in the Kitchen. Economic activity on Eleventh Avenue includes a street cleaner, draught driver, engine operator, and the "West Side cowboy" hired to precede the trains. (George Grantham Bain Collection, Library of Congress)

by multiple networked factors. And though they may have made their claims as exploited workers demonstrating labor solidarity, their positions as participants within a networked system of production similar to the longshore crew, and the increased acceleration of an already difficult work process, produced a contingent space of force and resistance. Labor solidarity in resistance to difficult and often exploitative working conditions was an uneven and often temporary condition, that rarely was static or long lasting.

Working, earning, making a living, socializing, honing skills, making connections, being mobile, and facing uncertainty—all were part of the working world of Hell's Kitchen. Conditions of uncertainty, such as fluctuations in the market, injury, new opportunities, or changes in domestic circumstances such as marriage and childbirth, resulted in complex and complicated strategies of labor and employment for many Hell's Kitchen residents. Spaces of work and labor were produced when the boundaries of economic activity overlapped, creating not economic scales but shifting sites of complex movement in a market system prone to breakdown, error, and misperception. Though the private economic marketplace clearly functioned to maximize profit and accumulation, its operation at the physical level was far from rational or efficient. The relationship between individuals and families attempting to make a living by earning a decent wage and securing the benefits offered by a changing economy and private employers determined to maximize return was mediated by spatial actors outside of the direct control of the powerful and less powerful. These spatial actors, including but not limited to the heat and noise of the laundry factory, the degraded physical state of piers and docks, the relative proximity of opportunities to earn, and the spaces of social networking, operated not merely as the capitalist spatial context in which the laboring (or resting) body was embedded, but as a network itself, producing the lived, conceived, and representational spaces of daily life. In other words, as the struggle between capital and labor produced work space, the work space itself, whether the ship's hold, the laundry room, or the factory floor, in turn produced the work-space relationship.

In a process similar to the restructuring of community space, detailed in chapter 4, economic space did not mask, conceal, or simply reproduce existing social and economic relations of exploitation but created conditions for creative or critical movements among the participants. For urban areas such as the Middle West Side, survival and accumulation strategies were carried out within the horizontally overlapping sites of production, distribution, and consumption, as well as differently moving sites that included, as active participants, rotting wood, rivers and oceans, clothing, politicians, capitalist owners of businesses, and the brick and mortar of home and work. All can be considered as forces of production. In this view, there is no one system that can be understood fully as an "economy," by moving back and forth between the local scales and larger, vertically situated scales. If all actants, the human and non-human alike, are considered in examining Hell's Kitchen's social relations, the question of interest takes on a quite different meaning. The rotting wood of the waterfront piers contributes to how the space of longshore labor is understood, and weighs as a factor in the epistemological categories, such as laboring men that Barnes and others utilize not only to describe the working population, but to act as advocates for improving what they see as inefficient and exploitative conditions. It contributes to both outsider human perceptions of the laboring community and to the attitude of those who labor toward their work, their self-identity, and their place in the wider social system. Yet the wood itself is produced by the system of unregulated investment and infrastructure development decried by Barnes and Tompkins and by a conceived natural process that cannot be separated from the process of human production. The very physical conditions that affect the social relations of the area, and other scales, are not produced by the capitalist market alone. The flow of the North River, the weight of cargo and human traffic, the riverine microbes that inhabit the wood, and the conditions of the wood's origins and value-added production into a pier, are only some of the factors within the connected network of actors that produce a working pier and its parts. It can also be stated that whatever long-term trajectory the wood follows cannot be understood within the epistemological categories used to understand human interest, yet the wood remains part of the network that produces the spaces of social reproduction.

One of the challenges when analyzing urban spatial processes, particularly where concerned with issues of labor, earning, employment, and survival, is to view these sites of production, distribution, and consumption as taking place within both the spatial matrix of capital circulation, and as part of networked systems whose linkages, connections, and simultaneous operations escape what philosopher Bruno Latour refers to as "the tyranny of distance and proximity." Latour suggests that "geographical proximity," the science of mapping, measuring, and triangulating, creates a limiting view of connectivity that can only be analyzed in terms of economic and social scale, and thus confined within the processes of capital accumulation and circulation.[296] What I suggest is not ignoring the obviously important process by which the demands and needs of accumulation and circulation create working space, but to think beyond the confines of labor geography and analyze capitalist urban space as embedded with a network of associations produced by human and non-human actors, and examine the ways that these approaches can combine to inform our understanding of the production of urban space.

Historians and sociologists of work and labor have made valuable progress in the field of labor geography, moving beyond a static view of class and social reproduction to develop sophisticated approaches to questions of work and space, particularly within the urban context.[297] Avoiding what Andrew Herod terms the "mechanistic theorization" of the production of space under capitalism, labor geographers present issues of class as simultaneously lived, imposed, mediated, and performed, but always under conditions dominated by the needs of capital. This view of class as changeable, ambiguous, and often based as much on performance and perception as on empirical conditions of accumulation allows for an analysis of laboring spaces that is in excess of the totalizing, or mechanistic, view prevalent in world-systems theories of *longue-durée* capitalist development.[298] The ambiguous nature of class and other identities available under capitalist production is neatly summed up in anthropologist Don Kalb's description of laboring space: "Capitalism in the abstract is a social process that continuously creates and re-creates concrete, specific, and friction-prone linkages between particular places, and shapes idiosyncratic and contradictory ties between disparate human groups."[299] Kalb points out that capitalist processes are "about timing and geography as much as about exploitation and control." The timing and geography of capitalist processes, the location of a node of production, the position of a given commodity within a chain of raw material, value-added labor, and consumption are adequate but limited starting points for the analysis of the production of working space that frames human life chances and explains workers' interests in making the geography of capitalism in particular ways.[300]

Work in labor geography has demonstrated how laboring populations often seek their own spatial fixes that are coincident with but sometimes opposed to the interests favored by capital and other groups. How these populations, such as skilled, unskilled, and semiskilled workers in Hell's Kitchen, view themselves as a population and as individuals within a system of production is dependent on their collective and singular attitudes regarding interests in the system itself. As Ira Katznelson and others have shown, the patterning of class in urban populations is not confined to the workplace division of labor and its attendant patterns of skill, wage, gender, ethnic, or racial divides, but is determined as well by other spaces, such as the home, and by institutional affiliations in community groups, associations, and informal understandings of interest.[301] But just as in the example of the wood that makes up a working pier, the question of interest is neither simple nor readily apparent when considering how space is produced through interactive networks that are often overlapping but simultaneously disinterested. Thus if we assume, as Herod suggests, that wage workers "have implemented in the landscape quite different sets of spatial relations" than those that accumulate and circulate capital, we further assume that the binary of worker-capitalist, no matter how nuanced the approach, serves as the prime producer of urban space. Though indeed less mechanistic than earlier models of the creation of urban space, such an approach nevertheless ignores or discounts the indeterminate set of interests, acting as forces that produce not neatly layered vertical scales, but points of assembly, composition, decomposition, movement, blockage, and coagulation that are better understood as horizontal.

Horizontal networks of organic, inorganic, and social materials can better be understood as sites of potential than as scales of the possible, when considering working piers or the human aggregate populations that interact with them.[302] Viewing the composition of such sites, and their decomposition, in no way ignores or obscures the social conditions that contribute to the process, nor negates the obvious fact that certain social relations, such as that of wage worker to employer, are reproduced within such sites. What such an approach does is examine what exists that determines and produces the social in ways that account for the multiple timings and geographies that overlay capitalist relations. If, as Kalb states, capitalism is about timing and geography, it is also, as a social system of relations, embedded within and overlapped at its porous boundaries, by temporalities that have no interest that can be discerned, and are outside of its purview. The resulting spatiotemporal conditions, the composition and decomposition of contingent sets of related actants, produce spatial sites as emergent properties that are only partially explainable or attributable to human social relations, be they dominated by capital or some other form of exchange and mediation.

In their article "Geography without Scale," authors Sallie Marston, John Paul Jones, and Keith Woodward posit emergent sites as "singular compositions rarely resembling discrete and unitary objects," which produce social space. In their formulation, sites of emergence are not scalar relations that can be charted through the connective impulses of geography, the "limited connectivity"[303]

Latour decries, but take place within temporally demarcated events. Sites include the "divers intermesh of languages and desires; the making of connections between bits of bodies and parts of objects; sentences half caught, laws enforced prejudicially, or broken accidentally." Thus sites, as objects of empirical study, require a "rigorous particularism" that acknowledges and captures the emergent nature of sites as human and non-human interaction. Such an approach to urban history, urban sociology, and spatial theory vastly complicates both the nature of the object and the epistemological categories used to interrogate them. Though this approach does not need to abandon the study of working populations or the geography of social relations within a capitalist framework, it does suggest that the ontological conditions of the urban, of space, and of history require a different set of tools. It also calls into question the relative adequacy of such categories as labor, community, workplace, and economic scale in interpreting the production of urban space.

The difficulty in moving beyond the static methods of thinking about labor, space, and the market are apparent when researchers turn from theory to the empirical in studying and describing working spaces. Too many proponents of post-social construction approaches to critical geography and urban processes fall back on reliance upon neologistic pronouncements about flows, mobility, and open-ended possibility. Often, trying to determine the actual physical makeup and affectual production of such sites involves examining so many horizontal assemblages that analysis of the kind provided by human social geography is not feasible, and thus requires a different understanding of spatial affect. But an examination of two particular laboring spaces, the ship's hold and the Higgins Carpet factory floor, lend credence to the concept of site as a valuable descriptive and analytical category for examining the urban work process.

In both instances, the multiple causal actants creating the sites of emergence combine different temporalities in the produced space, and produce often-incommensurate levels of interest. Tracing every network of organic, inorganic, and social materials that coagulate to produce the laboring conditions in either instance would require a book-length study to fully comprehend, and only then would the contingent nature of each site be brought into temporary epistemological frame. But in both instances what could appear as an economic process of labor exploitation, even accounting for the variety of levels of mediation, becomes the site of emergence when examined utilizing the post-social construction categories of critical geography such as *site*. As more historians explore, for example, the role of heat in contributing to the production of civil unrest, and sociologists examine how technological changes in communication affect international labor coalitions, we will be able to come to a better understanding of how non-human material factors contribute to the contingency of human affairs.

CHAPTER SIX

Hell, Death, and Urban Politics

On the other hand, most of the constructive programs of the present are the work of cranks and quacks. Reform has even come to be a matter of commercial exploitation. Whatever may be the value of exposure of abuses by cheap magazines and sensational newspapers, there can be no claim that they represent either expert knowledge of fundamental social principles or efficient skill in constructive reform.

—PROFESSOR U. G. WEATHERLY,
American Journal of Sociology, 1911

We're tired of Fifth Avenue telling Tenth Avenue how to live.
—WEST SIDE RESIDENT, 1913

"THERE IS A FINE OPPORTUNITY for an important public improvement in making Eleventh Avenue not only trackless, but a beautiful roadway, helpful to all residents of the nearby unattractive streets," states *New York Times* letter writer Anni Gould in the spring of 1911. Arguing for a change in perception of the area, she goes on, "The quarter has long been known as Hell's Kitchen. Why not let this neglected waterfront, so full of delightful possibilities, have its share of the generous plans enjoyed, for example, in the Bronx. It certainly would make better citizens if, instead of noise and dirt and an absolute lack of any point of interest, the people could enjoy a tree-lined boulevard, with parks, model tenements, playgrounds, a Carnegie library, garden plots and perhaps a kindergarten and nursery."[304] Mrs. Gould, a resident of Tarrytown, a New York City suburb, might have been accused of never having visited the area. By 1911, Hell's Kitchen, though not enjoying "tree-lined boulevards," had been physically restructured, including a park, recreation pier, new tenements, a public garden, increases in street repair and sanitation services, and a Rockefeller settlement house that included a community library.

An outsider, Mrs. Gould's protest on behalf of Hell's Kitchen was part of a general trend to reshape the perception of the area that had been ongoing for quite some time, carried out mainly by local residents and community leaders. One example, from the *New York Times* in 1900, written by Father Mooney, the parish priest of Sacred Heart Church at 42nd Street and Ninth Avenue, objects to the *Times*'s use of "Hell's Kitchen" in a report on new settlement houses:

> Permit me to protest emphatically against this wanton use of a disgraceful appellation. I know that it has been glibly applied at times to various sections of the west side in the flippant manner that reflects no credit on those who have done so. The people of this section, which is the latest to receive the opprobrious epithet, justly resent this slur cast upon them. Surely everything cannot be augured for the success of the movement itself, whose beginning is thus marked by an insult to those whom it professes to benefit. I would state, moreover, that many of the lauded features it proposes to introduce into the social and, impliedly [sic], religious, life of this neighborhood already exist there. Within the radius of 500 feet of this site (the new settlement house) are two schools giving religious education to more than 2,500 students, a kindergarten, a boy's club, a gymnasium, a young men's club—each club having it's [sic] own house and a young woman's club, which has been in existence for eight years, and in which, at various times, lessons in music, singing, dressmaking, cooking, physical culture and elocution have been given. The philanthropy, therefore, which inspired the movement in question will not find in the scene of its proposed operations the arid waste which the writer of the article in question evidently wishes to have regarded as the actual condition of this locality.[305]

Hell, Death, and Urban Politics

Freight train rumbling though Hell's Kitchen as children play on sidewalk.

Father Mooney's protest is clearly intended to defend his own position and work in the parish, and the vital nature of Sacred Heart Church as an institution important to the community. But in his umbrage, he expresses much about the residents of Hell's Kitchen, their wants and desires, and indeed, about restructuring and geographic citizenship. Far from being Balzac's "people without wants," Hell's Kitchen residents had many wants and desires, often the same wants as their middle-class neighbors to the north and east, and the reformers who took such an interest in their neighborhoods. Within the changing spatial structure of their area, Hell's Kitchen residents developed many wants, expressed often as political demands or claims, but also expressed as residents of a particular urban spatial location. In other words, to paraphrase Marx, in the face of spatial restructuring, Hell's Kitchen residents made their own spatial wants and desires, but they did not do so under conditions of their own choosing. Father Mooney expresses not only the fact of an existing community but also the self-reflexive knowledge of residents regarding how they are perceived by outsiders, who, according to Mooney, "wish" that the conditions of squalor that they assign to the area were the actual conditions. The desire articulated in Mooney's protest is to be taken seriously, to be part of the wider political community, welcomed in not as an "urban problem," observed through the veiled gaze of reformers and politicians, but accepted as fully functioning members whose values and wants, though perhaps at times at odds with the dominant normative claims of some segments of the community of citizens, were also often in concert with them. Residents of Hell's Kitchen were demanding their own particular form of visibility. They demanded that people who lived outside the area see them for what they were, not just as objects of study, sympathy, or derision. Though expressed differently, Hell's Kitchen residents' view of proper citizenship involved the same tensions that framed citizenship among other aggregate populations, and included both the demands for "rights" and the expression of responsibilities.

In this chapter I sketch the changes taking place among Hell's Kitchen residents during this period of spatial restructuring and tie together some of the major themes already addressed, focusing also on the wider picture of spatial change at various sites. I have shown in previous chapters the many ways in which space was both producer and produced in the Middle West Side, altering the ways people thought of themselves, and how outsiders perceived them, and the various factors that contributed to the production of such space. I broaden the scope of my investigation by interrogating the ways Hell's Kitchen residents were affected by networked projects of spatial restructuring that opened up new avenues of opportunity and mobility, while simultaneously closing other avenues and, in turn, altering the wants and desires of residents. What becomes clear is the many ways that the reformscape gaze and political and economic power in the city combined with non-human actants to alter the built environment, and thus altered the spatial horizon of residents, affectually framing the ways they viewed themselves as urban citizens. Spatial restructuring did not produce a monolithic bloc of new citizens, or serve as the container for socially constructed identities, but instead created a variety of contested perceptions of citizenship that produced spatial cognitions, causing some to become more mobile, some to cling to "traditional ways," and others to steer a middle path.[306] Spatial restructuring did not determine the outcome of individual action, but acted as a force of production, altering not only the physical landscape but also the perception of those who inhabited its lived space. Restructuring was not structurally determinate, but produced the opportunities to create sites of collective activity that affected how local residents and outsiders viewed Hell's Kitchen. Whatever path was chosen, the self-conceptions of these citizens was always contained, framed, and directed by the available spatial environment.

Urban theorist and philosopher Arjun Appadurai, speaking on contemporary forms of urban planning, suggests that what progressive planners can accomplish through working with local populations is the "designing of context" rather than the construction of "proper" environments for the urban poor. Appadurai makes the link between urban context and futurity, pointing to the ways that built environment shapes the thinking of residents through framing their aspirations, anticipations, and anxieties. Rather than planning cities based on the desires of landowners for anticipated profit, or to guard against the anxieties produced by the urban poor, Appadurai points to ways urban planners can work with underserved populations to provide the context for a future of positive possibilities.[307] I suggest here that what united the often disparate reformers we have termed "Progressive" was their very formation within the crucible of an urban spatial environment that was historically unique. Their attempts at tenement reform, park construction, settlement house education, and general uplift were all part of an effort to "design context," to create, in the words of social reformer Mary Simkhovitch, "the democratic—the bringing together people of different origins, education, and social opportunity."[308] In attempting to define, classify, and ameliorate the "urban problem," and to design the proper spatial context for good citizenship, urban reformers were produced by the very space in which they operated. As well, their conceptual vision of urban space produced a "population" of residents that reformers attempted to define, but who also defined themselves in the restructuring process.

THE RESTRUCTURED SPATIAL CONTAINER

IT WOULD BE DIFFICULT to list or describe the many instances of spatial restructuring occurring in New York City between 1884 and

1920 that affected the Middle West Side. It would be equally difficult to chart the variety of ways this restructuring changed the people who lived there. Here I will discuss several important changes and their spatial consequences, then examine how demands, wants, and desires were shaped and contained by the newly produced spaces of the city. In this process, spatial restructuring did serve to "open out" cityscapes, allowing for more physical mobility and creating different horizons of expectation, producing multiple networked connectivities that overlapped, produced change, and sometimes never came together.[309] But shifting physical spatial patterns at the same time produced different blockages and coagulations, creating sites of activity that patterned the spatiotemporal relations of the real and the virtual, of the now and the future, for both Hell's Kitchen residents and those outsiders whose opinions shaped the spatial perception of the area. This relationship, between the actually existing lived space and the conceived and representational, forms the crucial spatiotemporal nexus within which citizenship, democracy, and life chances, were contained.

Between 1894 and 1914, the population of the United States changed dramatically. In the well-told tale of immigration, the United States, for the first time, strictly regulated the immigration of non-white, non-European immigrants with the Chinese Exclusion Act of 1884, and opened the "golden doors" to immigrants from Europe. As immigration from the British Isles and northwest Europe continued, the bulk of "new" immigrants arrived from eastern and southern Europe. New immigrant groups included farm families and miners from the polyglot Habsburg Empire—Slavs, Hungarians, Croats, Poles, and Lithuanians—many pushed off their lands by high rates of taxation and the drop in commodity prices caused by American agricultural production. Many immigrants also arrived from southern Italy and Greece, from impoverished farming communities long dominated by oligarchic families. Jews from eastern Europe, escaping both poverty and persecution, came in large numbers. Though many immigrants eventually settled in Chicago, Cleveland, and other industrial cities, many arrived and stayed in New York, altering the existing structures of both the city at large and local areas.[310]

The effects of immigration on the Middle West Side were twofold. On the one hand, new arrivals, whether settling in Lower

Manhattan, Brooklyn, or elsewhere, created pressures on the existing residents by increasing the population that required inexpensive housing. Large numbers of ethnic groups often settled neighborhoods, creating enclaves that encouraged "outsiders" to seek other areas. This was particularly true of the German, Irish, and African Americans enclaves in Lower Manhattan, many of whom dispersed to the Upper East Side, the Middle West Side, Harlem, and eventually, the Bronx and Queens.[311] The shifting ethnic makeup of neighborhoods altered the physical terrain, with new stores, churches, social clubs, and restaurants taking over existing spaces, and new languages and customs appearing in everyday transactions. Ethnic solidarity, while never absolute, was an important aspect of daily life in procuring employment, housing, friendship, and leisure activities. Population pressures also led to more severe overcrowding, forcing ethnic populations to come to agreement over the use of spaces such as rooftops, streets, taverns, and play spaces. For the Middle West Side, ethnic tension over housing, control of streets, and basic values created tensions that while sometimes could turn violent, could just as often be creative.

Perhaps more important to the Middle West Side was the second aspect of immigration, the concerted attempts by "nativist" Americans to "assimilate" the new arrivals. With the sheer numbers of new arrivals causing increasing anxiety among the middle and upper classes, new attention, such as Riis's photo exhibits and the various committees formed to investigate vice and crime, was paid to the problems of congested neighborhoods. Though New York had long experienced the formation of slums, and calls for their reform or removal, the concentration of population brought about by new immigrants, combined with aging housing stock and limited housing options, made the period from 1894 to 1914 one where images of the "urban problem" proliferated with a multitude of consequences. Indeed, I would argue that the majority of spatial changes during this period can be linked to the urban anxieties brought on by immigration.[312] In essence, the spatial changes produced by the multiple networks that brought an influx of new arrivals to New York produced the Progressive urban reform movement.

Two of the more obvious consequences of immigration anxiety (besides the reform movement itself) were the tenement house improvement movement and the "Americanization" programs that flourished after 1900, both of which greatly affected the Middle West Side. The efforts to improve urban housing in New York City are framed by the first new tenement design competition in 1879 and the move to federal government involvement by the 1930s. The high point in New York was the 1901 Tenement House Committee, whose major recommendations became enforceable law, and whose actions coincided with the peak of overcrowding brought on by mass immigration. The commission's rulings divided the working-class New York neighborhoods between Old Law and New Law buildings, and opened the door to real city involvement in housing planning and eventual code enforcement.

Americanization, though concerned with immigrants, was applied to populations in the Middle West Side that were not composed of the newly arrived. Although manifestations of Americanization, from settlement houses to language schools and industrial training, were more prevalent on the immigrant-rich Lower East Side, the Kitchen certainly experienced its share of attempts to mold proper Americans. The establishment of an industrial school for boys sought to create productive workers from West Side youth. Hygiene and home-skill classes attempted to provide proper training for a middle-class domestic interior and for proper lifestyles. Carried out in schools and settlement houses, these programs were often causes of resentment from parents and targets of humor and sly derision from area children.[313] Ultimately, the settlement houses and training schools of the Americanization movement produced space that Middle West Side residents used, or did not use, to reinforce existing values and norms as well as forge new and different ideas about the meaning of citizenship.

One of the main consequences of immigration for the Middle West Side dwellers was the very creation of the "urban problem" and the creation of poor communities as objects of observation and reform. In the mid-nineteenth century, slum populations were typically considered morally degraded and hopeless, a segment of society to be isolated, avoided, and ignored whenever possible. Moral failing was located at the level of the individual, and the space created to encourage moral turpitude was the creation of the individuals themselves. By 1910, slum populations, now dominated in many neighborhoods by new immigrants, had become the objects of study of social workers and spatially oriented reformers, those attracted to theories that linked proper space to proper citizenship.[314] The consequences of this repositioning of the urban poor had profound ramifications for the residents of the Middle West Side, and for other and subsequent urban populations. Indeed, two important developments in the formation and production of the urban problem begin in the 1880s, with the establishment of the first urban social science specializations in U.S. universities, and the beginnings of the settlement house movement, a widespread attempt to apply the scientific findings of social science to a constructed clientele. The concepts of community and place, cornerstones of decades of urban development projects ostensibly designed to "empower" urban dwellers, were produced from these contingent spatial conditions.

Transportation changes were another aspect of spatial restructuring that altered life in the city, and perceptions on the Middle West Side. The congested conditions begin in the 1880s after the completion, in 1879, of the Ninth Avenue elevated steam line. The expansion of streetcar service, horse-drawn, then electric, in the 1880s and 1890s, brought more population to the region while incrementally expanding opportunities for employment for some. For many of the poorer residents, riding the train or streetcar was often a luxury, though the demands of employers for working hours often restricted workers to living near their place of employment. The street and train cars were also spaces where the city's population, at least those below the socioeconomic level of the private carriage, would mix in public, providing a semi-democratic experience of the city.[315]

The construction of the city subway system, beginning with the IRT line in 1904, brought even more change to physical mobility and mental perception, opening various areas of the city to everyday travel. Mobility meant that employment opportunities not previously available broadened the economic horizons of some workers, allowing prospective employees to work farther from home.[316] Yet all of the changes in city transportation infrastructure occurred within and through other spatialized networks of connection and disconnection that sometimes overlapped or operated in isolation to form the more localized changes.[317] While many New York area residents took advantage of the new opportunities afforded by changes in the transport system, Hell's Kitchen residents faced frustration in their efforts to affect physical changes. In two cases discussed in this chapter, Hell's Kitchen residents sought the removal of two dangerous segments of the city's transport infrastructure, only to be denied due to political and economic structures beyond their immediate control.

The transportation revolution caused by the subway brought new opportunities in the form of leisure choices and housing location. As leisure time expanded for some segments of the working classes, novel forms of leisure activity, connected to the developing culture of consumption, attracted the attention of Hell's Kitchen residents. Two new and developing American forms, movies and the amusement park, shaped leisure time activity, and presented the urban population with new vistas of wants and desires. Historian Kathy Piess, in her groundbreaking work on gendered leisure activities at the turn of the century, documents the various ways that young working women produced spaces for the expression of a new freedom centered on fashion, consumption, and aggressive sexuality.[318] The Middle West Side social surveys document the importance of dance halls, "nickel dumps" movie houses, and other sites of public amusement to area residents, particularly young working women. In her study *The Neglected Girl*, Ruth True reported that dance halls were the location of girls seeking excitement before "keeping steady company," whereas the motion picture theaters, often converted storefronts, were the haunt of "tough" girls "lacking the tawdry finery" of girls with factory jobs.[319]

Consolidated New York City was home to one of the earliest efforts at constructing "amusement parks" for the middle and working class. Coney Island's various incarnations, including Luna Park, Steeplechase Park, and Astroland, were pioneering efforts in providing working-class urban audiences with affordable entertainment that attracted people, especially the young, through forms of presentation that altered the time-space perspective of the daily routine. While the new amusements were designed to produce spaces of fantasy and entertainment for lower- and middle-income New Yorkers, many of the physical spaces designed for such play and recreation mirrored the very industrial-technological processes in which most workers spent the majority of their laboring time.[320] So while new opportunities for public amusement were available, sociability and interaction were still largely played out in the interstitial spaces of Hell's Kitchen—the alleys, rooftops, street corners,

Hell, Death, and Urban Politics

The worst elevated train disaster in NYC history. Sept. 11, 1905, the 9th Ave. El derails at "suicide turn" in Hell's Kitchen. (Milstein Division of United States History, The New York Public Library, Astor, Lenox and Tilden Foundations)

and hallways that had long been the location for such public forms of contention, negotiation, and conviviality.

On an everyday level, the transportation revolution brought other spatial changes, on the streets and in the air. The new popularity of the automobile brought the "motorcar" through the Middle West Side, often with unwanted consequences. Before the implementation of traffic signals and the stricter enforcement of speed regulation, autos were often the cause of tragic accidents. Residents frequently complained of wealthy motorists speeding through their streets, causing havoc. A 1905 account of a trial reports how Hell's Kitchen residents "ganged up" on a motorist who had accidentally run over a young neighborhood boy, testifying that he acted with disregard and impunity, and deserved a jail sentence for reckless manslaughter.[321]

A further change was the replacement of horsepower by the new machines. Our contemporary sensibilities can only imagine the difference made in everyday city living by the slow disappearance of the veritable army of animal power that made up a major part of the city's economic engine. Horses were utilized for public transport in horse-drawn rail cars, and also served as power for the wealthy in their private coaches. They were also the power behind moving vans, trash collection, food and ice deliveries, and provided employment for trolley drivers, coachmen, teamsters, and those involved with general horse maintenance, such as cleaning the waste the animals left in their wake. Horses were a constant and ubiquitous presence on the Middle West Side, often dying on the streets and becoming objects of fascination for young children. Young girls formed rings from the hairs either collected from or yanked from horses' tails, and particularly prized the tail hairs of "golden" or red horses, which could be skillfully woven into plain brown hairs to form fashionable rings and bracelets.[322]

With reform both a reality and an often-empty slogan, city politics shifted from the 1880s to the 1920s, and contributed to spatial restructuring. As the concept of reform shifted from moral uplift to the more scientific approach, city government and service agencies changed in scope and operation. Reform and fusion tickets for local office shifted political loyalties, as politicians and their supporters attempted to break traditional party loyalties, often by presenting themselves as nonpartisan and scientific, reflecting the period's rising emphasis on notions of expertise in public administration. By 1900, with certain exceptions, most mayoral campaigns claimed some part of the title of reform, and backed up their public campaigns with expert testimony and studies from quasi-governmental research groups, like the New York Bureau of Municipal Research, which were proliferating. While Tammany Hall Democrats retained influence, often openly campaigning against the "goo-goo" outsiders on anti-reform platforms,[323] reformist politics and politicians played an important role in altering the urban landscape.

For the Middle West Side, reform politics and political party alignments tested and shifted loyalties and voting behaviors. The election of Thomas McManus, founder of the McManus Democratic Club, to the New York State Senate in 1898 came at the expense of Tammany stalwart George Plunkitt, who prided himself on being able to deliver the working-class vote through extensive patronage. Though many Middle West Side ethnic and social clubs supported Tammany-influenced Democrats, reform candidates such as August Belmont drew considerable support from the area. Reform meant the patronizing attitudes of "outsiders," a view that many Tammany

Democrats played to their advantage. Along with a distrust of reform politicians, many West Side residents were equally wary of social workers connected with the elite Columbia University and with the settlement houses that dotted Manhattan.

In their proposal for an independent school of social science in 1918, the group of scholars who would eventually found the New School for Social Research famously referred to New York City as "the greatest social science laboratory in the world." Charles Beard, one of the leading figures of the group, declared that "every problem occurring in the social world" was evident in New York.[324] Though Beard's pronouncement must be taken as the hyperbolic claim of a zealot inspired by his own position within a thriving academic community, his statement reflected the increasingly prevalent processes of urbanization in the nation as a whole, and particularly the Northeast. New York indeed can be taken as the crucible for many reform programs, and its reform community, centered in the United Charities Building of the Charity Organization Society on 22nd Street, was a leader in the national movement, but the urban problem and its relation to space was evident in many American cities of the period. The combination of reform politics, Progressivism, and the increasing importance of social science research altered both the physical landscape of Hell's Kitchen and the perception of the residents themselves.

Father Mooney's protest—his demand that outsiders, social reformers, reporters, and city officials take his district seriously as part of the city's political community—arises in the midst of the changes illuminated above. If New York was indeed the crucible of reform, and the immigrant community of the Lower East Side was under the reform microscope, the Middle West Side and its spatially perceived monotony and degradation was the very embodiment of urban space as the modern urban problem. If New York was the crucible, the place where social and economic forces cause elements to change, the results were the production of space that in turn produced the very terms and knowledge by which that space was understood, evaluated, and altered. What was produced was the urban problem itself, the reform movement meant to restructure it, and the population of residents, whose wants and desires were formed within the crucible as well.

SELF-PERCEPTION AND CITIZENSHIP

IN 1908, A NEW YORK reporter related a story, no doubt embellished for the entertainment of his readers, that is indicative of the process of the spatial production of desire, a direct representation of the ways in which outlook and potential are altered by new experiences of space. Responding to the problem of repeated attacks on motorcars by young men in Hell's Kitchen, the members of the Big Brothers movement hit on a solution. They gathered groups of young "ruffians" from the neighborhood and took them for rides in the automobiles, demonstrating the power and freedom such technology can bring. The children were enlightened, and quickly became auto enthusiasts, troubling the drivers of vehicles no more.[325] Given the proclivity of newspapers like the *New York Times* to use Hell's Kitchen stories as forms of textual slumming for its readership, the story must not be taken at face value. It is doubtful that each child emerged from this experience with a new sense of appreciation for the automobile and immediately commenced plans to become a member of the class of people who could afford such luxury. But there is evidence that the spatial restructuring of the Middle West Side did affect the attitudes and outlooks of some residents, sometimes resulting in organized political demands and other times resulting in individual decisions. As changes in transportation, employment, and housing infrastructure presented new opportunities, some chose to leave, others to stay; some took part in contestations over the use of new spaces, and some attempted to alter the very language used to evaluate the spatial environment.

The derogatory moniker "Hell's Kitchen" was not the only widespread negative nickname to plague the Middle West Side. It was joined, in just two instances, by "Abattoir Road" to delineate the section of Eleventh Avenue and West 39th Street bordering the series of slaughterhouses, and the infamous "Avenue of Death," the well-known cautionary nickname for the stretch of the same street on which a freight line ran through the mixed residential-industrial district. It is probable that many localisms for other sites in the area have escaped the historian's gaze, but one thing is certain from the historical record: Hell's Kitchen residents, especially the younger ones, were very aware of their neighborhood's reputation and the multiple meanings attached. It is also clear that by 1905 at least some residents were joining Father Mooney's protest, demanding that outsiders no longer use the offending nickname. As well, disputes over such issues as the use of settlement houses and the opening of a recreation pier coincided with the establishment of several local community clubs and an increased political presence in the area. The participation of Hell's Kitchen residents in these spatial contestations does not indicate that they suddenly discovered their class interests, but it does show in the historical record an increased awareness of their position as objects of sociological study, and an increased desire for the rights of full citizenship.

Spatial change on the Middle West Side accelerated after 1900, with changes in tenement laws and the construction of DeWitt Clinton Park. In both instances, the needs of local residents triumphed, at least temporarily, over the accumulation needs of private property. Whereas the tenement laws were largely imposed from above, with little citizen participation, the impetus for the park came in part from the demands of local residents, and their decision to back an anti-Tammany Democrat in the 1900 New York State Senate race. The legislation that assured the financing of DeWitt Clinton Park had been sponsored by Thomas McManus, whose McManus Democratic Club had been established in 1898 to both counter the power of other city clubs and to serve the increasing needs of Hell's Kitchen. McManus had won election to the State Assembly over the Tammany Hall candidate, and though he seems to have served his own interests as a real estate developer well, he also was influential in bringing the grievances of Hell's Kitchen residents to public attention. His election coincided with a citywide scandal over the manipulation of ice prices, an issue vital to Middle West Side residents. When it was shown that the "ice trusts" were in collusion with several top-ranking Tammany politicians, many former supporters turned their votes to "reform" Democrats like McManus.[326] McManus would eventually make peace with Tammany, but he was still seen by some residents as their best hope for local improvements. While in the Senate, McManus also sponsored bills to create the adult education Evening School on Tenth Avenue, and promoted legislation to increase dock facilities and improve waterfront maintenance. McManus's election over Tammany stalwart George Plunkitt is but one indication of the changing attitudes of Hell's Kitchen residents during the period.

Subsequently, other issues of spatial restructuring were raised, with some resulting in improvements and others falling victim to private interests or legislative inertia. Of the three issues discussed here, one concerned mainly city authorities and two dealt with strictly local issues, yet all three were in part the result of the demands of local citizens. One long-running issue, the removal of freight tracks from Eleventh Avenue, the infamous "Avenue of Death," would not be resolved until the 1930s, despite repeated efforts by local residents and politicians. Conversely, disputes over the opening of a recreation pier and the reuse and redeployment of a local settlement house were issues settled by the combined efforts of locals and social workers. In all three instances, spatial conditions created changing wants and desires that coincided with ongoing disputes between residents, reformers, and city authorities.

The "Avenue of Death," Eleventh Avenue, maintained dangerous conditions for Hell's Kitchen residents. With the advent of steam locomotives in the 1860s, the slow-moving horse-drawn trains were replaced by a fast-moving menace, particularly in the area between 23rd Street and Riverside Park. Though no accurate statistics exist for death and injury totals, opponents of the rail line claimed up to one hundred people per year were killed or injured by the trains, many in Hell's Kitchen. The property of the Central and Hudson Rail Company, the line brought freight from Albany to the city, often for shipment from the New York piers to their ultimate destination. The large locomotives were equipped with hand brakes operated by a single brakeman from the top of the car, and stopping depended on his skill and alertness, as the trains often reached speeds of 25 miles per hour. As early as 1866, a state senator referred to the operation of freight trains on street surfaces as "an evil which has already endured too long."[327] The only concessions made by the Central Hudson to public safety was to pay mounted flagmen, usually from outside of the city, to ride in front of the trains warning of their approach.[328] The riders became a favorite sight among Hell's Kitchen youth and were often the target of rock throwing, and were often blamed for lax performance at the scene of an accident. The fight against the rail line, at a peak between 1906 and 1914, would linger on until the 1930s, when it was finally replaced by a series of compromise structures, including the West Side high line.

Hell, Death, and Urban Politics

Tracks on Death Avenue. (Milstein Division of United States History, The New York Public Library, Astor, Lenox and Tilden Foundations)

The fight over removal of the tracks had a long history, but no serious action had ever been proposed before 1900. A report to the Board of Estimate by city engineer Ernest C. Moore expressed the frustrations and difficulties in considering proposals for removal. Moore acknowledged that nearly forty years of proposed plans had amounted to stalemate, as the land had been the private property of the Central Hudson Rail Company, and its commerce important to city businesses and employees. Moore points out that the rock layer along the avenue made a submerged line unfeasible, and the cost of connecting the trains to docks through short elevated lines was financially prohibitive. In spite of "aesthetic objections," Moore recommended a four-track elevated line running from lower Manhattan to Riverside Park, a plan that would not be adopted for another twenty years.[329] So though the issue for most local residents was the clear danger imposed by the freight trains moving through the center of their street, the process of their actual removal involved legal questions of property rights, the right of the city to seize it for the public good, the costs of redirecting the lines, and the economic impact on both suppliers and consumers of freight goods.

Pressure to improve general conditions in areas like Hell's Kitchen combined with a number of local incidents to cause an increase in activism against the rail line between 1906 and 1911.[330] Though no reliable statistics on accidents involving pedestrians exist, the increase in the area's population and the increased visibility of poor neighborhoods due to urban reformers brought more attention to the issue after 1900. However, it does appear that the push for city authority to force the Central Hudson Company to remove the tracks through condemnation originated in pressure brought by N.Y. State Senator Martin Saxe, a representative of the powerful Upper West Side and associate of the West End Association, which wanted the tracks adjacent to Riverside Park removed. Saxe credited the occurrence of Hell's Kitchen fatalities with swaying his fellow senators to support his 1906 legislation, which granted the city the right to demand removal or move toward condemnation.[331] According to Saxe's account, the transportation committee of the State Assembly was in a late meeting on the very issue when word of the gruesome death of another young person crossing the tracks altered the terms of the debate. Saxe reports that on the evening the bill was debated by the Cities Committee of the N.Y. State Senate, news of the death, on 30th Street and Eleventh Avenue, swayed his fellow committee members to vote the bill to the floor, where it passed by majority vote. Saxe's bill allowed the city to proceed with condemnation of the New York Central's property on Eleventh Avenue if no compromise plan could be worked out.

In the face of charges and counterconcharges, the two sides launched a battle of legal rights and publicity, as track opponents held public meetings and supporters of the railroad publicized the cost of track removal and the threat to property rights. A variety of plans were proposed for the freight line by all interested parties, including the Rapid Transit Commission, the Central Railroad Company, and local civic groups such as the West End Association and the West Side Taxpayers Association. The dispute centered around three key issues: what would replace the Eleventh Avenue line, who would pay for it, and how would the city be compensated for additional land grants to New York Central for the necessary changes. By March of 1907, proceedings had become so contentious that the *New York Times* reported that the situation was "a hopeless muddle" that would no doubt wind up in court where it would "drag on for

Hell, Death, and Urban Politics

years."[332] As the May 1908 deadline of the Saxe bill for condemnation passed, no compromise seemed imminent as neither side nor the city administration would agree to a comprehensive plan for track removal, and the New York Central challenged the legality of the Saxe legislation.[333]

As the city headed for a legal showdown with the rail company, a particularly gruesome "Death Avenue" incident made the headlines. On September 25, 1908, seven-year-old Seth Low Hascamp, named for a former mayor of the city, was crossing Eleventh Avenue on his way to St. Ambrose School when a freight train hit him, "tearing his small body in half." By witness accounts, Hascamp had tried to pass over the connecting apparatus of a stopped train, followed by a group of friends. When the train started forward, the boy was thrown to the tracks and cut in two. Though it was illegal for trains to stop on Eleventh Avenue to avoid just such incidents, a coroner's inquest exonerated the engineer and his crew and placed the blame on Hascamp's "own negligence." The decision by city coroner Shrady on October 25 ignited a wave of protest in Hell's Kitchen, as groups of children paraded up Eleventh Avenue, banging makeshift drums and lighting firecrackers. They were led by Henry Shroeder, secretary of the Track Removal Association, who claimed that 198 children had been killed in a ten-year span, many in the late afternoon in winter, when visibility was difficult, and many children crossed the tracks while they were "carrying dinners to their fathers at work."[334]

The organized protest continued for several days after the coroner's decision was made public, and Shroeder and his organization vowed to continue the public demonstrations until "this dangerous hazard is finally removed." The march appeared to be spontaneous and the cause of track removal was popular in the area. In a canny tactic, Shroeder utilized the image of the deserving poor, merely bringing dinner to hardworking fathers, in an appeal to the wider public for action. The protest action continued for several weeks, eventually drawing smaller crowds of marchers. But the incident served to bring more local residents into the fight and altered the tactics of the opposition, who changed their name from the less-threatening Track Removal Association to the "League to End Death Avenue." Shroeder and other activists emphasized the "protection of schoolchildren" in their campaigns, although many adults were killed or injured by the trains, and made concerted efforts to involve more Hell's Kitchen residents in the protests. "We are determined to make use of the indignation which has been aroused among the dwellers in this neighborhood to obtain removal of the deadly tracks."[335]

Community involvement in the track removal fight temporarily spiked in the aftermath of Hascamp's death and residents of Hell's Kitchen continued to complain about the dangers of the tracks, yet removal proceeded slowly, according to the timetables of the courts, the city administration, and the New York Central. By 1911, although there was still local involvement, the League to End Death Avenue had become a vehicle for the Central Federated Union interested in preventing the New York Central's monopoly over rail shipping on New York's West Side and across the river in New Jersey. The battle over removal, spurred to public interest by deaths in Hell's Kitchen, also shifted by 1911 to concerns over Riverside Park and the aesthetics of any new plan for construction. In a 1916 public meeting, Board of Aldermen president Frank Dowling suggested that the League to End Death Avenue had been misnamed and was merely representing the interests of local freight workers and residents living near Riverside Park. "The League wants the tracks removed, but it does not want the Central to 'stick up the city' as the price of removal." At the same public meeting, Charles W. Staughton, president of the Municipal Art Commission, demanded that any changes be placed under the auspices of the Committee for Improving Riverside Park, and that "artistic considerations" be given top priority.[336] The debate had come a long way from Seth Low Hascamp.

While the battle over track removal was clearly fought at the level of economic and legal interests unconcerned with the hazards of Eleventh Avenue in Hell's Kitchen, the dangerous conditions provided local residents with sites to perform political acts expressing their demands for improved conditions. The protests of October 1908, following the death of Seth Hascamp, provided young Hell's Kitchen residents with an opportunity to display their genuine displeasure with existing conditions and a disciplined approach to public protest that clearly impressed news reporters attending the events. In an area where local journalists often went to great lengths to describe locals in "colorful" and often demeaning terms, the news accounts all

Hell, Death, and Urban Politics

noted the solemnity, good order, and determination of schoolchildren marching with American flags draped in mourning bunting.

While the concerns of local residents were largely denied in the disputes over public railways, two more localized issues combined the changing nature of the area with an increased demand among residents that they be allowed more input in decision-making processes. The establishment and maintenance of a city-financed amusement pier on at 51st Street in 1901 and the "seizure" of an abandoned Rockefeller-financed settlement house illustrate these spatial productions of want and desire. In the case of the Settlement House, the contentious relationship between "Fifth Avenue and Tenth Avenue" was brought into sharp focus. As Rockefeller, dissatisfied with the results of settlement house work, withdrew his financial support, local residents, dissatisfied with the often-condescending attitude they perceived in settlement house workers, seized the building for their own purposes. In the struggle over the maintenance of an amusement pier, Hell's Kitchen residents demanded the right to entertainment and relaxation within their own space, and access to forms of space available to other city residents. In both cases, local spatial concerns overlapped with other spatial networks to produce sites of political performance.

The announcement in October 1900 by philanthropist John D. Rockefeller of the establishment of a new settlement house for Middle West Side residents was greeted with enthusiasm by the city's reform community. The new settlement, located at Tenth Avenue and 50th Street, was to be five stories of limestone and buff brick, intended to "meet the needs of the neighborhood in educational, social, and other ways." The Settlement House, as it was dubbed, was to be equipped with "public baths, a manual training school, a cooking school, libraries, classrooms, club rooms, and a general assembly room," according to its manager, A. A. Hills. Hills had been hired by Rockefeller and his supporters at the Fifth Avenue Baptist Church due to his success at running settlement houses in Lexington, Kentucky. Mr. Hills stated that it was his objective to make the house act "in perfect sympathy" with the local neighborhood, and turn it into the "most popular institution" in Hell's Kitchen.[337]

The Settlement House joined several other philanthropic institutions in Hell's Kitchen dedicated to providing services to local residents. The Fifth Avenue Baptist Church had, several years before, established "The Armitage" on West 47th Street and Eighth Avenue, providing day care, kindergarten, and "Americanization" classes to locals. In 1897, the Association for Improving the Condition of the Poor had helped establish Hartley House at West 46th Street to "teach the laboring people to make home attractive" as well as providing a free library and work for destitute women. The settlement houses were to serve the dual purpose of providing charity services to the deserving poor and working to "Americanize" immigrants and the children of urban workers.[338] At the opening of Hartley House on 1897, reformer R. Fulton Cutting stated, "When we can teach the poor to keep their homes attractive, we are striking at the root of poverty."[339]

Social settlement houses and the efforts to "uplift" the deserving urban poor produced coalitions of wealthy philanthropists, religious leaders, and social scientists, whose efforts and conceptions of their "clients" produced contentious urban spaces. Ruth True's work with West Side children and adolescents shows their general hostility and distrust of the settlement houses and those who ran them. She reports that no activity could go forward without the tacit approval of the "tough kids," who refused basic training in hygiene and deportment, and constantly disrupted meetings and classes. She further reports that many of the young people were told by their parents not to frequent the settlement house.[340]

In January of 1910, after reconsidering his commitment, John D. Rockefeller and the Fifth Avenue Baptist Church removed their financial support from the Settlement House. Citing a perceived "lack of real progress" in the work of the house, Rockefeller and the Charity Organization Society announced that all financial support would be withdrawn as of January 15, and that the house was to be vacated within four days. The announcement stunned the settlement workers, some of whom had worked at the house for nine years, and local residents. The settlement workers immediately announced that they would seek other funding and a buyer for the property. But two years later, the house stood mostly unused, with showings of films and the occasional use of the basement gym to stage boxing matches.[341]

By 1912, the old Settlement House building had become part of a larger battle over local control and participation, along with the fight over the use of Hell's Kitchen's only recreation pier. Established in 1900 and funded by the city's Department of Docks and Ferries, the amusement pier was equipped with electric lighting, a refreshment stand, musical performers, and a candy stand. Covered and fairly well maintained, the pier provided space for Hell's Kitchen residents to socialize near the cooler air of the river. Like other New York recreation piers, the 50th Street Pier was franchised to business interests who ran the refreshment stands and "floating pools" in the North River for profit, and the pier opened in early June and closed in September when children returned to school.[342] Though it provided a needed and desired service to the community, the pier was not under local control, and did not serve as a focal point for political organizing of any kind. The pier had been seized by the city from private owners for lack of upkeep, and converted into an amusement pier in 1899. Its location was less than ideal, one pier downriver from a condemned pier formerly owned by George Plunkitt, and just upriver from two working piers and a trash dump.[343] Nevertheless, the piers were popular and well used, in spite of the offal smells and the presence of criminal elements along the waterfront.[344]

Whereas many Hell's Kitchen residents were deeply suspicious of the settlement houses, and the mission of the houses was the cause of many disputes among locals, amusement piers providing safe and accessible entertainment were generally agreed to be a positive addition to the area. Despite the criminal presence along the piers, the recreation pier was apparently well policed, and no alcohol was allowed. Recreation piers and "floating" pools had, by 1900, become very popular attractions for city residents living near the waterfronts.[345] One point of contention was whether these sites of recreation should be provided by the city, privately owned and operated, or some combination of public-private partnership. By 1912, with the closing of the Rockefeller Settlement House, the recreation pier at 50th Street had taken on added importance, becoming more than merely a location to escape the heat and grind of the crowded tenements.

In March 1912 a group of Hell's Kitchen residents, including local merchants, businessmen, youth groups, men's social clubs, and several social workers, devised a plan to change not only the space of Hell's Kitchen, but its perception to outsiders. The plan included year-round use of the 50th Street pier, and the reopening of the Settlement House. The *New York Times*, covering the story with the headline "Hell's Kitchen Will Reform Itself," initially reported that former settlement house workers from the defunct Rockefeller project devised the plan. The paper was forced to print a retraction two days after the original story when it was learned that the plan had indeed come from a coalition of local groups, and was mainly supported by local youth "gangs," whose representatives informed the reporter that they would no longer accept their neighborhood being referred to as "Hell's Kitchen," a name which, according to a *Times* reporter, all local residents oppose "with great exception."[346] The coalition of groups, which included the Raleigh Athletic Club, the Go-Aheads, the Liberty Athletic Club, and the Yankee Doodle Youths, staged a public meeting at P.S. 17, at 47th Street between Eighth and Ninth Avenue, to air their plan and garner widespread public backing.

The plan included demanding public funding for year-round maintenance and access to the amusement pier, and for the reopening of the former Settlement House building, with both structures being managed and operated by local organizations, and meeting the needs of local citizens. The meeting was so well attended, especially by school-age children, that many could not gain entry. By the end of the public meeting, attended by State Senator McManus, Recreation Commission Secretary Bascom Johnson, and members of the New York City Board of Education, local residents were openly expressing their collective demands for city services, public funding, and, most importantly, respect. As the first mass meeting ever sanctioned by the Board of Education in a Middle West Side public school, the affair became a community event, with a moving picture shown before the meeting, and a band supplying music before and after the proceedings.[347]

Much of the meeting's content—gleaned from newspaper reports, as no official record exists—consisted of criticisms of previous reform efforts and attempts to reconcile the views of "outsiders" with the needs of local citizens. But the main thrust appears to have been demands that local individuals and organizations be allowed complete autonomy in deciding how the two spaces, the pier and the former Settlement House, would be utilized. The promoters of the plan to reuse both spaces intended that they serve as "social rendezvous" points for all clubs within the district, and that each organization be allowed to contribute ideas as to how the pier and settlement building should be organized. "This population knows how to take care of itself," promoters were quoted as saying, as they proceeded into a thinly veiled critique of impositions and paternalism by philanthropists and outsiders. Though acknowledging that all local settlement houses and their workers provided valuable and useful services, the Rockefeller House had failed because "Fifth Avenue had tried to tell Tenth Avenue how to live."[348]

Father Mooney's protest against the paternalism and dismissive nature of outside perception in 1900 echoed a growing sentiment among area residents, and outsiders as well.[349] Though Mooney was not the first to publicly decry the appellation "Hell's Kitchen," his protest came at a specific, contingent moment in the history of the Middle West Side, one structured in part by the spatial changes occurring in the area. As ideas about urban reform shifted from moral uplift to spatially based "scientific" reform, the urban population of the Middle West Side became, for outsiders, a discrete object of social study, and the spaces in which they lived produced, in part, urban progressivism itself, in its many forms. An increasing self-awareness of their position as social objects and the provision of spatial restructuring projects in their lived environment altered the relationship between residents and reformers, city authorities, and one another. The dual demands, that outsiders drop the use of Hell's Kitchen and that Fifth Avenue cease telling Tenth Avenue how to live, are the most obvious examples. Other examples include public meetings held by local business owners demanding expanded police presence, and demands for more park space to supplement DeWitt Clinton Park.[350] Spatial restructuring, such as the placement of the amusement pier, the locating of settlement houses, and the long struggle over freight tracks, produced not a unified class-consciousness, nor an entrenchment of particularized spatial defense, but rather the various coalescent sites of spatial production that shifted and altered the very language used to evaluate its use. It exemplifies the essence of urban spatial production.

The struggles over Death Avenue, the Rockefeller House, and the recreation pier illustrate several points in understanding both the history of the Middle West Side and urban studies more generally. All three examples of contention and conflict over existing urban space and its potential restructuring were examples of the process of urban spatial production formed in the nexus of lived, conceived, and representational space. The lived space of the Middle West Side, its conception in the plans of reformers, landlords, politicians, and residents, and the representational experience of spatial restructuring produced the space for multiple, lasting forms of knowledge and urban understanding. Accurate or not, the altered spaces produced concepts of urban reform, planning, population control, community, and place both particular to the area and generalized in urban theory. As Lefebvre suggests, the ability of restructured produced space to reproduce existing social relations is key to the production of urban space under capitalist modes of production.[351] In other words, how representational space is structured by conceptual space determines the measure, and the limits, of social reproduction. It follows that when spatial change, mediated by other social changes, does not produce the same code, then the behavior, attitudes, and outlook of those utilizing the lived space can be altered and affected.

The three examples of contention also illustrate the relational nature of urban spatial production, and the multiple ways that urban space structures meaning within epistemological systems. Along with Father Mooney's demands that the use of "Hell's Kitchen" as a moniker cease, demands for spatial change reflected residents' concern over the aesthetics of the area, as well as its perceptual reputation. This acknowledgment that the language used to evaluate the spatial conditions of the area was in part produced by the space itself demonstrates the relational nature of spatial production. The urban space of the Middle West Side was not produced in isolation, as a discrete site constructing static identities. Multiple connective networks, physical as well as cognitive, produced relational spaces that, though capable of reproducing existing social relations, were also capable of altering them. The relationship between the local Irish-American population and the Irish "old country" is just the most glaring example of this process.

As lived, conceived, representational, and relational space, these three sites of potential are all affected by non-human actants forming networked connections across scales and perceived boundaries. Each site produces not a predetermined result of structured space but the potential for altering human understanding in ways that are outside of the designs of powerful forces that attempt to reproduce favorable social relations. Rather than producing the closed system desired by those concerned with social control, the restructuring and alteration of space, affected by so many factors, operates as an open system that, though able to coagulate and produce blockages that do reproduce existing relations, can take a variety of potential directions, leading to contention, struggle, and altered political, economic, and social relations.

The results of these tripartite productions of urban space were neither the creation of place and community nor the formation of neatly understandable political and social formations. Instead, the constant process of spatial production *produced* the very language through which humans evaluated their life chances and positions within differently arrayed and connected economic and social systems. Thus the very nature of the ever-shifting process of interaction between human perception and actual physical space makes producing a "snapshot" of an existing, static community impossible.

CONCLUSION

The Spatial Production of Desire

FOR RESIDENTS OF HELL'S KITCHEN between 1894 and 1914, wants changed in part based on spatial restructuring. The spatial changes occurring at the local, city, and regional level altered the basic relationship of the horizon of futurity, the relation between the immediate and the virtual horizon of the possible. What takes place in this historical setting can neither be described as the building of community nor as the creation of place. Community, as far as it can be delineated, always existed in the residential population, and was always in the process of formation, reformation, and disintegration.[352] Place, as understood by Yi Fu Tuan and other human geographers as a source of confidence, a space navigated as home, proves an inadequate definition of the shifting terrain experienced by residents.[353] In fact, even the concept of shifting, overlapping, and overdetermined scale, though utilized in this study, seems to fall flat when considering the complexities of historical development in Hell's Kitchen or in other similarly situated urban areas. To fully consider what "takes place" in the spatial productions described, and determine what is produced, what is constitutive, and what is contained, I return to several questions raised in the opening chapter, and reconsider them in light of the evidence, both theoretical and empirical. In so doing, I conclude with four main points first approached in the introductory chapter, and explored through the historical evidence from the Middle West Side between 1894 and 1914. They are:

- That urban space is produced in a constant process shifting between lived, conceived, and representational space.
- That all produced urban space is relational, neither simplistically located nor utilized, but produced as a set of relations.
- That the production of urban space is ultimately the product of the interactions of contingent network connections that include both human and non-human actants.
- That because of their connection to various theoretical and empirical projects, common urban spatial conceptual categories such as community, place, and scale, though useful, are ultimately inadequate for forming a deeper, more complex understanding of historical and current urban spatial processes.

The major purpose of this undertaking was to examine the historical relationship between spatial restructuring and citizenship and, as the title suggests, determine the ways that "democracy" is spatially contained in a given historical period and setting. Democracy is used here to indicate both the Merriam-Webster's definition of the term, as in "rule by the people," and a set of practices related to social institutions that allow for the practice of group empowerment. In other words, democracy is the legal right to participate in the governing system and the practical means to access justice. In the legal sense, Hell's Kitchen residents, by 1919,[354] had achieved the official means for participatory democracy for all naturalized adults. They possessed, in legal terms, the same rights as other American citizens, to vote, assemble peacefully, engage in public discourse, petition for redress of grievances, and defend themselves in open courts.[355] What is more interesting, and harder to define, are the multiple ways "rights" work at the level of everyday life, and the relationship between rights and space. One can possess every legal right of citizenship, and still be denied the basic security and privilege that comes with the actual practice of the relationship between citizen and public and private authority. What I have tried to show in this study is how citizenship, in part the right to have demands taken seriously, is spatially determined, and that the ability to have impact on questions of social justice is both embedded in and intertwined with the urban spatial environment.

To return to David Harvey's "difficult prospect" that the "standards of social justice" are constituted in what he terms "socio-ecological processes," is to reexamine the process of the production of urban space. Harvey suggests that how we understand citizenship, in terms of both legal and informal practice, is spatially determined. In the former, spatial demarcation of inclusion and exclusion is basic to the practice of citizenship.[356] In the latter, how we think of a particular population, demarcated spatially and socioeconomically, determines how seriously we take their political demands and responsibilities.[357] Here again, space is not the context of action, but a co-constitutive element in the construction of social reality. Spatial location, spatial perception, and spatial imaginings—the lived, conceived, and representational of spatial production—constitute or create the standards by which not only the present and past but the future are judged. For the Middle West Side, the physical environment constitutes the language of reform, the perception of space frames the impunity of policing, and the representational spaces change wants and desires. Indeed, for Barnes's study of dockworkers, the enormous silence he discovers in academic writing on longshore workers is attributed to the spatial perception of the men as "unworthy of study."[358]

For Harvey, the spatial determinism of the language of evaluation is bound up with the needs of capitalist accumulation in the industrial and post-industrial city. In this formulation, the needs of local residents, as workers and as citizens, are subsumed under the needs of the system of capital investment, production, circulation, and profit. Through complex mediation, the needs of accumulation, based on the "spatial fix," set the parameters, or framework, in which discussions and negotiations over land use, investment, improvement, and restructuring are held within the dominant discourse of accumulation. This spatially determined discourse thus reproduces workers according to the needs of a machine-like system of accumulation and circulation, which Harvey labels the "production of spatial difference," which is the key mechanism of uneven geographic development.[359] Though a powerful argument regarding the processes of urban development and a valuable tool for both historical and contemporary analysis, uneven geographic development relies upon an economic determinism that is inadequate for examining all urban assemblages, and falls short of supplying a textured account of what is produced within the urban nexus.

Capital accumulation strategies structure much of the spatial environment of modern cities, and were certainly influential in the production of space in New York City between 1894 and 1914. But though they are influential, they are not determinate. In any site of accumulation, the space is itself constituent of the strategy. For example, Kansas City, as an inland city without ports, does not attempt to attract deepwater shipping. More complexly, labor market radicalism, city government, proximity to markets, and numerous other factors produce space that either fits accumulation strategies or must be altered, if possible. Space entrepreneurs must conceive of the space as a potential site of accumulation before deciding to invest. Conceived space is not simply determined by accumulation needs, but is a constant process of interaction with the lived and representational space of all actors, both human and non-human.

The modern industrial city is an accumulation strategy that produces space, but it also an archive of information, a layered warehouse of tectonic knowledge, a point in a system of movement, and the site of multiple strategies linked to a variety of potentials. It is ultimately run by no one nor controlled by a single logic or institutional design. Nor is it the "organic" creation of multiple individuals. Rather, it is a combination of all these and more, of multiple modern processes that, like Wolgang Shivelbush's steam rail engine, are dialectically bound to their own form of accident. Shivelbush argues that the creations of modern technologies carry within them their own potential for accident. The larger and more encompassing the machine, the greater the potential damage.[360] In the American case, the city presents the site of all potentiality: for Jefferson, it is the location where democracy is impossible; for reformer Frederick Howe, its only real hope, the always already of utopia and dystopia. Though cities are contingently rational, flowing, controlled, and contained, they are contiguously blocked, clogged, and coagulated, often outside the control of even the most powerful human coalitions, and are spaces of potential error.

Understanding the city as the site of potential error, and what this means for the production of urban space, requires thinking about urban populations and their actions in a different light. Rather than view the city or the urban region as scaled locations operating within relatively efficient systems of production, distribution, and consumption, we might want to see the urban, in light of the empirical evidence, as the always-in-process, contingently produced sites of relationality, with as much potential for error and misrecognition as for efficient production. Here, no point of urban space is ever a discrete unit socially created by multiple actors for definitive purposes. In an unpublished paper on urban potentialities, Adams and Enigkbo suggest "critical movement" as one way of understanding what bodies do and perform in sites of social reproduction, misperception, and potential error.[361] Critical movements can explain both unplanned actions to negotiate the everyday lived space of the city and coalescent collective actions that occur in sites of spatial production. For the Middle West Side during the period under study, critical movements can include the shifting loyalties of class, ethnicity, race, and gender exemplified by police rioting and changing labor conditions, as well as more organized collective actions such as the attempted redeployment of the recreation pier and settlement house. What I have tried to show through historical evidence is that the production of urban space, though often unduly influenced by powerful spatial coalitions, is the product of a variety of factors that often exceed the ability to control it, and is affected by numerous overlapping networks of forces that do not socially construct "reality" but produce sites for often incommensurable actions. It is in these produced spatial sites that the very language of evaluation, of social justice, is lived, conceived, and represented.

THE FUTURE OF CITIES

THE SPATIAL PROCESSES of urbanization between 1894 and 1914 changed the role and perception of urban areas in American culture. Industrialization, immigration, urban consolidation, political reform, and spatial restructuring produced and were products of the changes wrought by urbanization. Whereas historically contingent ideas of the role of cities were formulated, such as Frederick Howe's "hope of democracy," the intimate relationship between urban areas and American economic, social, and political development was firmly established. Nostalgia for a sylvan America of open plains, small towns, and small-*r* republican values would and do linger, and suburbanization, globalization, and regionalism have altered perceptions of the role of cities, yet they remain bound up in the physical functioning of the nation and in the formation of theories and opinions regarding the future of the nation.[362]

How are we to understand the historic role of the city in American life, and in questions regarding "global" cities and capitalist development? How does the history of Hell's Kitchen, one small, relatively insignificant segment of one very important and influential city, inform a different understanding of urban space, the urban past, present and future? Is it enough, as Edward Soja urged twenty years ago, to "take space seriously" as a category of social analysis?

Soja's groundbreaking 1989 work, *Postmodern Geographies*, urged urban theorists to move from viewing space as the physical context of human social interaction to examining spatial production as, following Lefebvre, both "producer and produced" in human aggregate populations.[363] Soja's call for a spatialized ontology urged urbanists to understand space as a "social product," that is, "simultaneously the medium and outcome, presupposition and embodiment" of human interaction, a "second nature" that "transforms both physical and psychological spaces." Soja's work, along with translations of Lefebvre and the efforts of human geographers like David Harvey, has altered many of the terms of the urban debate, and "taking space seriously" is no longer a new and fresh approach to urban studies but an accepted method employed by numerous scholars, planners, politicians, developers, and community-based organizations. The "spatial turn" in human geography and other fields has produced a disparate body of knowledge concerning spatial production, and urban space particularly, and has advanced both social theory and social practice regarding urban development.[364]

Yet twenty years on, the spatial turn for many urban historians, theorists, and sociologists remains contained within the limited frameworks of the dialectical explanations of spatial production promoted by Soja, Harvey, and many others. The spatial turn has produced theories and descriptions ranging from the romantic urban visions of Marshall Berman and Jane Jacobs[365] to the catastrophic urban nightmares of Mike Davis. The city is the potential space of political regeneration, as in the promotion of the "right to the city," or the spatial location of neoliberal "revanchism," where gentrified urban yuppies displace "authentic" populations of working-class communities.[366] Urban theorists continue to describe space as the context of human action, and many remain locked into static notions of place, community, and scale that inhibit rather than promote new thinking regarding questions of urban history and urban development.

With notable exceptions, taking space seriously has resulted in valuable work, but work that continues to locate urban space as context, and continues to engage in rearguard actions that react to powerful forces of development and accumulation, rather than anticipate the way actual space is produced, and in so doing, encourage potentials that are open. This book on New York's Hell's Kitchen insists that, both historically and in contemporary theory, actual physical space must be included as constituent in understanding the contingent compositions that alter urban space and spatial practice. If, as I have suggested, sites are the contingent locations where human and non-human actants form coagulant moments of alteration, both producing space and being produced by it, and all productive forces must be taken into account in any attempt at description, analysis, or policy recommendation, then several implications arise for urban history and urban theory.[367] First, as no account can accurately represent the vast multitude of factors that act, then static and predictive notions of history and theory must be inadequate. Thus viewing urban space as historically fixed at any given period, relying on notions of place and community, must produce only partial and incomplete histories of urban spatial processes. Likewise, these same variant factors make a working synthesis of theory and practice in urban politics impractical and indeed impossible. There is no perfect fit between the universal and particular, between theory and practice, that is adequate as either description or proscription, nor should achieving such a synthesis be the goal of urban theory.

So does a contingent analysis of urban space preclude a progressive politics and resign urban theory to endless modes of description, abandoning the actual city to the workings of the dominant modes of accumulation? To answer, I return to Henri Lefebvre and his ideas regarding the production of space. Lefebvre's relationship to the Situationists has been well documented, as has his split with Guy Debord and others over issues of theory and practice. Although he certainly held the work of the Situationists in high esteem, and admired their attempts to remake urban space, he ultimately rejected their performative "play" as mere reaction to the processes of capitalist reproduction. Lefebvre opted for an urban theory grounded in history, but a history not contained by ideology. Lefebvre's urban history, as with his urban theory, remained painstakingly open to the processes of productive forces, a view grounded in the physical realities of human and non-human interaction. "If space is produced," he states,

> if there is a productive process, then we are dealing with *history*. The history of space, of its production qua reality, and of its forms and representations, is not to be confused either with the causal chain of historical events, or with a sequence, whether teleological or not, of customs and laws, ideals and ideology, and socioeconomic structures or institutions.[368]

For Lefebvre, as for myself, space is not context, not simply the location of human action, but a real actor with its own history, one that participates in other temporal formations, with or without what we can understand as an "interest." The history of space, and the history and process of urban development, thus affected, must remain radically open to potential, even under conditions of oppression, exploitation, and accumulation. As Lefebvre states,

> Some over-systematic thinkers oscillate between loud denunciations of capitalism and the bourgeoisie and their repressive institutions on the one hand, and fascination and unrestrained admiration on the other. They make society complete; they thus bestow a cohesiveness it utterly lacks upon a totality which is in fact open, so decidedly open, indeed, that it must rely on violence to endure. The position of these systematizers is in any case self-contradictory: even if their claims had some validity they would be reduced to nonsense by the fact that the terms and concepts used to define the system must necessarily be mere tools of the system itself.[369]

Bibliography

Primary Sources

BOOKS AND MANUSCRIPTS

Allen, William. *Universal Training for Citizenship and Public Service.* New York: Macmillan, 1919.

Anthony, Katherine Susan. *Women Who Must Earn*, West Side Studies series. New York: Survey Associates, Russell Sage Foundation, 1914.

Baker, Abigail Gunn. *Municipal Government of the City of New York.* New York: Ginn and Company, 1916.

Barnes, Charles. *The Longshoremen.* Russell Sage Foundation, Survey Associates 1913.

Bruere, Henry. *New City Government.* New York: D. Appleton, 1913.

Cartwright, Otho Grandford. *The Middle West Side: A Historical Sketch.* New York: Russell Sage Foundation, 1914.

Cleveland, Frederick. *Efficient Citizenship.* New York: Longman and Green, 1913.

Flagg, Ernest. "The Plan of New York and How to Improve It." *Scribner's Magazine* 36, August 1904.

Fuller, William B. *The Proposed Filtration of the Croton Water Supply of New York City.* New York: Municipal Engineers of New York, 1908.

Goldmark, Pauline. *West Side Studies: Boyhood and Lawlessness.* College Park, MD: McGrath Publishing, 1914.

Herzfeld, Elsa. *Family Monographs: The History of 24 Families Living in New York's Middle West Side.* New York: Kempster Printing, 1905.

Howe, Frederick C. *The City: The Hope of Democracy.* New York: Charles Scribner and Sons, 1905.

Marsh, Benjamin. "City Planning in Justice to the Working Population." *Charities and the Commons,* 19 February 1908.

———. "Economic Aspects of City Planning." *Proceedings of the Municipal Engineers of New York,* Paper No. 57, 1910.

Olmsted, John C. "The Relation of the City Engineer to Public Parks." *Journal of the Association of Engineering Societies,* 13 October 1894
Patten, Simon. *The New Basis of Civilization.* New York: Longman and Green, 1906.
Phelps Stokes, I. N. *The Iconography of Manhattan Island 1498–1909.* 9 vols. New York: Dodd Publishing, 1921.
Riis, Jacob. *How the Other Half Lives.* New York: Dover, 1971.
Shurtleff, Flavel. "The English Town Planning Act of 1909," in *Proceedings of the Second National Conference on City Planning.* Boston: National Conference on City Planning, 1910.
Smith, Robert A. C. *The West Side Improvement and Its Relation to All of the Commerce of the Port of New York.* New York: M. B. Brown Printing and Binding, 1916.
Taylor, Frederick Winslow. *The Principles of Scientific Management.* New York: Norton Publishing, 1911.
True, Ruth. *The Neglected Girl.* Russell Sage Foundation, 1908.
Veiller, Lawrence, and Robert Deforest. *The Tenement House Problem.* Macmillan and Sons, 1903.

GOVERNMENT DOCUMENTS

Board of City Magistrates. *Yearly Report City of New York Magistrates Court First Division.* 1908.
Department of Docks and Ferries for the City of New York. *Annual Report.* 1900–1920.
Board of Estimate and Apportionment Committee on Port and Terminal Facilities.
Report of the Committeee on Port and Terminal Facilities. New York: M.B.
Monthly Review of Labor Statistics. Washington, D.C.: Government Printing Office, 1910–1918.
National Civic Federation. *Report on the Proceedings of the Conferences on Immigration, 1906–07.* National Civic Federation Immigration Department.
New York City Magistrate Court. *People vs. Arthur Harris.* October 29, 1900.
New York City Police Department. *Annual Reports.* 1895–1910. New York State Assembly. *Proceedings.* Vols. 3 & 4, 1898.
New York Department of Taxation and Assessment. *Land Value Maps of the City of New York.* New York: Department of Taxation, 1890–1909.
New York City Parks Department. *Annual Report, 1902, Report on the Maintenance of Parks.* New York: Department of Parks, 1913–14.
New York State Public Service Commission. "First District Case Files."
New York State Court of Appeals. *Cordigan vs. City of New York* Supplemental Volume 34.3.
New York State Legislature. *Report of the Special Committee Appointed to Investigate the Police Department of the City of New York.* January 18, 1895.
Report to the Board of Estimate and Apportion on the Removal of 11th Avenue Tracks. December 1910.

Report of the Special Committee Appointed to Investigate the Police Department of the City of New York. January 18, 1895. In *New York City Police Corruption Investigation Commissions, 1894–1994.* Edited by Gabriel J. Chin. Buffalo, NY: W. S. Hein, 1997.

West End Association. *Proposed Plans and Agreement between New York Central & Hudson River Railroad and the City of New York.* New York: City of New York West End Association, 1913.

―――――. *Report of the Committee on Legislation, Law and Schools.* New York: City of New York West End Association.

―――――. *Report to the City of New York, 1899* West End Association, 1900.

ORGANIZATIONAL PUBLICATIONS

Bureau of Public Health and Hygiene. *Comfort Stations in New York City.* New York: Association for the Improvement of the Condition of the Poor, 1908.

Bureau of Municipal Research. *Municipal Reform and the Citizen.* New York: Bureau of Municipal Research, 1908.

―――――. *Some Results and Limitations of Financial Control.* New York: Bureau of Municipal Research, 1908.

―――――. *Report on the Methods of Supplies and Repairs of the New York City Police.* New York: Bureau of Municipal Research, 1908.

―――――. *Six Years of Municipal Research for the City of New York.* New York: Bureau of Municipal Research, 1911.

Citizens Union of the City of New York. *Report on the Proposed Relocation of the NewYork Central Railroad Tracks upon the West Side of Manhattan Island.* New York: Citizens Union of New York City, 1916.

Patriots League of America. *City Problems.* New York: Patriot League, 1909.

Real Estate Record and Builders Guide. Vols. 65, 67, 70, 83, 90. New York: C. W. Sweet.

NEWSPAPERS

New York Times
New York Press
Irish Times
New York World
New York Tribune
New York Herald
New York Call
Daily Forward

CENSUS DATA

U.S. Bureau of the Census. *Eleventh Census of the United States, 1890.* Vol. 1. Washington, D.C.: Government Printing Office, 1890. *Twelfth Census of the United States, 1900.* Vol. 1. Washington, D.C.: Government Printing Office, 1900.

Secondary Sources

Arato, Andrew, and Jean Cohen. *Civil Society and Political Theory.* Cambridge, MA: MIT Press, 1992.
Bachin, Robin F. *Building the South Side: Urban Space and Civic Culture in Chicago, 1890–1914.* Chicago: University of Chicago Press, 2004.
Banta, Martha. *Taylorized Lives: Narrative Structure in the Age of Veblen, Ford and Taylor.* Chicago: University of Chicago Press, 1993.
Barrett, James R. *Work and Community in the Jungle: Chicago's Packinghouse Workers, 1894–1922.* Chicago: University of Illinois Press, 1990.
Beauregard, Robert. *Voices of Decline: The Postwar Fate of U.S. Cities.* Routledge Press, 2003.
Beckert, Sven. *The Monied Metropolis: New York City and the Consolidation of the American Bourgeoisie, 1850–1896.* Cambridge: Cambridge University Press, 2001.
Bourdieu, Pierre. *Outline of a Theory of Practice.* Cambridge: Cambridge University Press, 1977.
Boyarin, Jonathan. "Space, Time, and the Politics of Memory." In *Remapping Memory: The Politics of TimeSpace.* Minneapolis: University of Minnesota Press, 1994.
Boyer, M. Christine. *Dreaming the Rational City.* Boston: MIT Press, 1982.
Bremner, Robert. *From the Depths: The Discovery of Poverty in the United States.* New York University Press, 1956.
Burrows, Edwin G., and Mike Wallace. *Gotham: A History of New York to 1898.* New York: Oxford University Press, 1999.
Castells, Manuel. *The City and the Grass Roots.* Berkeley: University of California Press, 1984.
Chauncy, George. *Gay New York: Gender, Urban Culture, and the Making of the Gay World, 1890–1940.* Basic Books, 1994.
Chudacoff, Howard. *The Age of the Bachelor: Creating an American Subculture.* Princeton: Princeton University Press, 1999.
Crary, Jonathan. *Suspensions of Perception: Attention, Spectacle and Modern Culture.* Cambridge, MA: MIT Press, 1992.
Dahlberg, Jane. *The New York Bureau of Municipal Research.* New York: New York University Press, 1966.
Daniels, Roger. *Guarding the Golden Door.* New York: Hill and Wang, 2004.
Day, Jared N. *Urban Castles: Tenement Housing and Landlord Activism in New York City, 1890–1943.* New York: Columbia University Press, 1999.
Easterling, Keller. *Enduring Innocence: Global Architecture and Its Political Masquerades.* Cambridge, MA: MIT Press, 2004.

Bibliography 243

Faure, Alaine. "Local Life in Working-Class Paris at the End of the Nineteenth Century." *Journal of Urban History* 32/5 (July 2006): 761–72.
Fogelson, Robert. *Downtown: Its Rise and Fall.* New Haven: Yale University Press, 2001.
Foucault, Michel. *Technologies of the Self: A Seminar with Michel Foucault.* Amherst: University of Massachusetts Press, 1988.
Fraser, Nancy. *Justice Interruptus.* New York: Routledge Press.
Geist, Charles. *Empire Builders and Their Enemies.* New York: Oxford University Press, 2000.
Gilfoyle, Timothy. *City of Eros: New York City, Prostitution and the Commercialization of Sex.* New York: Basic Books, 1992
Green, Martin. *New York, 1913: The Armory Show and the Paterson Strike Pageant.* New York: Macmillan, 1995.
Habermas, Jurgen. *The Structural Transformation of the Public Sphere: An Inquiry into a Category of Bourgeois Society.* Cambridge, MA: MIT Press, 1999.
_____. *Theory of Communicative Action: Reason and Rationalization in Society.* Boston: Beacon Press, 1984.
Hammack, David. *Power and Society in Greater New York.*
Hartog, Hendrik. *Public Property and Private Power: The Corporation of the City of New York in American Law, 1730–1870.* Chapel Hill: University of North Carolina Press, 1983.
Harvey, David. *Spaces of Capital: Towards a Critical Geography.* New York: Routledge, 2001.
_____. *Spaces of Hope.* Oxford: Blackwell, 2000.
_____. *Justice, Nature and the Geography of Difference.* Oxford: Blackwell, 1996.
_____. *The Urban Experience.* Baltimore: Johns Hopkins University Press, 1989.
Hepp, John Henry. *The Middle-Class City: Transforming Space and Time in Philadelphia, 1876–1926.* Philadelphia: University of Pennsylvania Press, 2003.
Herod, Andrew. "Workers, Space and Labor Geography." *International Labor and Working Class History.* No.64 (Fall 2003): 112–138.
Hofstadter, Richard. *The Age of Reform: From Bryant to FDR.* New York: Verso, 1955.
Jackson, Kenneth. *Crabgrass Frontiers: The Suburbanization of America.* New York: Oxford University Press, 1985.
Jacobs, Meg. "Democracy's Third Estate: New Deal Politics and the Construction of a Consuming Public." *International Labor and Working Class History* 55 (Spring 1999): 27–51.
Johnson, Marilyn. *A History of Police Violence in New York City.* Boston: Beacon Press, 2004.
Kahn, Jonathan. *Budgeting Democracy: State Building and Citizenship in America, 1890–1928.* London: Cornell University Press, 1998.
_____. "Re-Representing the Government and Representing the People." *Journal of American History* 19/3: 84–103.
Katznelson, Ira. *City Trenches: Urban Politics and the Patterning of Class in the United States.* Chicago: University of Chicago Press, 1981.

Kern, Steven. *The Culture of Time and Space, 1880–1918*. Cambridge, MA: Harvard University Press, 1988.
Kisselkoff, Jeff. *You Must Remember This*. Baltimore: Johns Hopkins University Press, 1989.
Kleinberg, S. J. *The Shadow of the Mills: Working-Class Families in Pittsburgh, 1870–1907*. Pittsburgh: University of Pittsburgh Press, 1989.
Kraut, Bennie. *From Reform Judaism to Ethical Culture: The Religious Evolution of Felix Adler*. Hebrew Union College Press, 1979.
Lears, Jackson. *Fables of Abundance: A Cultural History of Advertising in America*. New York: Basic Books, 1994.
Lefebvre, Henri. *The Production of Space*. New York: Verso, 1974.
Lockwood, Charles. *Manhattan Moves Uptown: An Illustrated History*. New York: Barnes and Noble Books, 1976.
Logan, John, and Harvey Molotch. *Urban Fortunes: The Political Economy of Space*. Berkeley: University of California Press, 1987.
Lynch, Kevin. *The Image of the City*. Cambridge University Press, 1960.
Marston, Jones, and Woodward. "Human Geography without Scale." *Transactions of the Institute of British Geographers* 30 (2005): 416–32.
Massey, Doreen. *Spatial Divisions of Labour: Social Structures and the Geography of Production*. London: Macmillan, 1984.
Mattson, Kevin. *Creating a Democratic Republic: The Struggle for Urban Participatory Democracy during the Progressive Era*. University Park, PA: Penn State University Press, 1998.
Montgomery, David. *Workers' Control in America: Studies in the History of Work, Technology and Labor Struggles*. Cambridge: Cambridge University Press, 1979.
Nasaw, David. *Going Out: The Rise and Fall of Public Amusements*. New York: Basic Books, 1993.
Oestreicher, Richard Jules. *Solidarity and Fragmentation: Working People and Class Consciousness in Detroit, 1875–1900*. Chicago: University of Illinois Press, 1989.
O'Connor, Richard. *Hell's Kitchen: The Roaring Days of New York's Wild West Side*. New York: Lippincott Company, 1956.
Ossofsky, Gilbert. "Race Riot 1900: A Study of Ethnic Violence." *Journal of Negro Education* 32/1.
Otter, Chris. "Making Liberalism Durable: Vision and Civility in the Late Victorian City." *Social History* 27/1 (January 2002).
Page, Max. *The Creative Destruction of Manhattan, 1900–1940*. Chicago: University of Chicago Press, 1999.
Peiss, Kathy. *Cheap Amusements: Working Women and Leisure in Turn-of-the-Century New York*. Philadelphia: Temple University Press, 1986.
Plunz, Oliver. *A History of Housing in New York City*. New York: Columbia University Press, 1999.
Porter Benson, Susan. *Counter Cultures: Managers and Customers in American Department Stores, 1890–1940*. Urbana: University of Illinois Press, 1986.

Rao, Vyjayanthi. "Slum as Theory: The South Asian City and Globalization." *International Journal of Urban and Regional Research* 30/1 (March 2006): 225–32.

Revell, Keith. *Building Gotham: Civic Culture and Public Policy in New York, 1898–1938*. Baltimore: Johns Hopkins University Press, 2003.

Richardson, James F. *The New York Police: Colonial Times to 1901*. New York: Oxford University Press, 1970.

Ridge, John T. "Irish County Societies in New York, 1880–1914." In *The New York Irish*. Edited by Ronald Boyer and Timothy Meagher. Baltimore: Johns Hopkins University Press, 1996.

Rodgers, Daniel T. *Atlantic Crossings: Social Politics in a Progressive Age*. Cambridge, MA: Harvard University Press, 1998.

Roediger, David. *Working Toward Whiteness: How America's Immigrants Became White*. New York: Basic Books, 2005.

Rosenzweig, Roy. *Eight Hours for What We Will: Workers and Leisure in an Industrial City*. Cambridge: Cambridge University Press, 1983.

Rice, Bradley. *Progressive Cities: The Commission Government Movement, 1901–1920*. Austin: University of Texas Press, 1996.

Salins, Peter. *New York Unbound*. Maiden, MA: Blackwell, 1998.

Sayer, Andrew. "Behind the Locality Debate: Deconstructing Geography's Dualisms." *Environment and Planning* 23 (1991).

Schultz, Stanley. *Constructing the Urban Culture: American Cities and City Planning*. Philadelphia: Temple University Press, 1989.

Shivelbush, Wolfgang. *The Railway Journey: Rail Travel in the 19th Century*. New York: Verso, 1985.

Shklar, Judith. *American Citizenship: The Quest for Inclusion*. London: Harvard University Press, 1991.

Skoronek, Stephen. *Building a New State: The Expansion of Administrative Capacities*. Cambridge: Cambridge University Press, 1982.

Slater, Don. *Consumer Culture: Consumption and its Uses*. Cambridge University Press, 1999.

Smith, Neil. *Uneven Development: Nature, Capital, and the Production of Space*. Oxford University Press, 1990.

———. "New Globalism, New Urbanism: Gentrification as Global Urban Strategy." *Antipode* 6/2 (2002).

Smith, Rogers M. *Stories of Peoplehood: The Politics and Morals of Political Membership*. Cambridge: Canbridge University Press, 2003.

———. *Civic Ideals: Conflicting Visions of Citizenship in U.S. History*. New Haven: Yale University Press, 1997.

Soja, Edward. *Postmodern Geographies: The Reassertion of Space in Critical Social Theory*. New York: Verso, 1989.

Steinmetz, George. "Critical Realism in Historical Sociology." *Comparative Studies in Society and History* 40 (1998): 170–186.

Swyngedouw, Eric. "Neither Global nor Local: Globalization and the Politics of Scale." In *Spaces of Globalization: Reasserting the Power of the Local*. Edited by Kevin Cox. New York: Guiliford Press, 1997.

Talen, Emily. "Beyond the Front Porch: Regionalist Ideals in the New Urbanist Movement." *Journal of Planning History* 7/1 (February 2008): 20–47.

Teaford, Jon. *The Municipal Revolution in America*. Chicago: University of Chicago Press, 1975.
Tuoraine, Alain. *Critique of Modernity*. Oxford: Blackwell, 1995.
Varga, Joseph. "For Speaking Jewish in a Jewish Neighborhood: Civil Rights and Community–Police Relations during the Postwar Red Scare, 1919–1922." In *Uniform Behavior: Police Localism and National Politics*. Edited by Stacy McGoldrick and Andrea McArdle. New York: Palgrave Macmillan, 2006.
Vidich, Arthur. "The Moral, Economic and Political Status of Labor in American Society." *Social Research* 49/3 (1982): 752–90.
Zukin, Sharon. *The Cultures of Cities*. New York: Blackwell, 1995.
Weibe, Robert. *The Search for Order, 1877–1920*. New York: Hill and Wang, 1965

ELECTRONIC RESOURCES

Appadurai, Arjun. "Disjuncture and Difference in the Global Economy." http://www.intcul.tohoku.ac.jp/~holden/Mediated Society/Readings/2003_04/Appadurai.html.
Benjamin, Walter. "Critique of Violence." http://generation-online.org/p/fpagamben.htm.
Community Board 4–Clinton/Hell's Kitchen District, http://www.manhattanch4.org/.
Mitchell, Don. "The Liberalization of Free Speech." http://www.maxwell.syr.edu/geo/faculty_current/mitchell.htm.
Foucault, Michel. "Of Other Spaces." http://foucault.info/documents/heteroTopia/foucault.heteroTopia.en.html.
Kopel, David. "Crime: The Inner City Crisis." http://www.davidkopel.org/CJ/Mags/InnerCityCrisis.htm.

Notes

1. Henri Lefebvre, *The Production of Space,* trans. Donald Nicholson-Smith (Oxford: Blackwell Publishing,1991), 46. Emphasis in original.
2. Otho Cartwright, *The Middle West Side: A Historical Sketch* (New York: Survey Associates, 1912), 52. Cartwright's historical sketch was included in the West Side Studies series funded and carried out by the Russell Sage Foundation as the first large-scale study of the residents of the Middle West Side.
3. Ibid., 55.
4. This was a well-documented tendency of Progressives, illustrated early on by Jacob Riis, whose famous photographs of the "other half" were intentionally shot to emphasize the most squalid corners of slum life. Cartwright's co-workers in the West Side Studies series purposefully excluded the more materially successful residents of the Middle West Side in order to focus upon those who required assistance.
5. Stanley K. Schultz, *Constructing the Urban Culture: American Cities and City Planning, 1800–1920* (Philadelphia: Temple University Press), 1989.
6. The name Hell's Kitchen, by most accounts, derives from the fact that a gang called the Hell Boys allegedly fenced their stolen goods in the area, thus their "kitchen." Other stories include the alleged statement of a local police officer, who claimed, when asked if the area was like hell, that it was hotter than hell, more like hell's kitchen. The most innocent story is that the name derived from a German restaurant named Heils. Special thanks to Polly Bookout and her associates at the Clinton–Hell's Kitchen history website: http://home.earthlink.net/~pbookhout/chp.html#WEB.
7. Katherine Anthony, *Mothers Who Must Earn* (New York: Survey Associates, 1912),16. Anthony's study was part of the West Side Studies series. For Anthony, it is the "American atmosphere" of the larger city that produces the "little snobberies" of the poor.

8. Commonly and officially referred to as the North River. The appellation appears to survive from the period of Dutch settlement, and was used to delineate the lower Hudson from the East River between Manhattan and Brooklyn-Queens.
9. This demarcation is based on a variety of sources, including a contemporary website, and is the official demarcation for the Russell Sage West Side Studies series, which is referred to throughout this work. The boundaries are porous, and are extended in some accounts when it is convenient, particularly for serialized accounts of crime in the tabloid press. Conversely, certain accounts locate Hell's Kitchen as farther west than the Seventh Avenue border. But Cartwright, in his historical sketch, describes the entire West Side below Riverside Park (72nd Street) to the Battery as desolate and devoid of architectural and communal character.
10. It was the southernmost part of the larger district of Bloemendale, or Bloomingdale, after a Dutch suburb. Cartwright, 11.
11. Ibid., 7–8.
12. Ibid., 3.
13. Hendrik Hartog, *Public Property and Private Power: The Corporation of the City of New York in American Law, 1730 to 1870* (Chapel Hill: University of North Carolina Press, 1983); Jon Teaford, *The Municipal Revolution in America* (Chicago: University of Chicago Press, 1983).
14. Jared N. Day, *Urban Castles: Tenement Housing and Landlord Activism in New York City, 1890–1943* (New York: Columbia University Press, 1999). Day argues that the process of subleasing, in which building owners, often investment groups, sublease tenement buildings to individuals who must then turn their profit from the rents they collect, is the key to understanding the creation of "depressed" areas and the deterioration of housing stock.
15. Edward Soja, *Postmodern Geographies: The Reassertion of Space in Critical Social Theory* (New York: Verso, 1989).
16. The first official zoning ordinance in New York City was approved in 1916. Based on the recommendations of several commissions representing mainly real estate interests, the ordinance regulated mostly the height of buildings, but also included restrictions on use depending upon area. But as many scholars of zoning point out, with the first zoning ordinances came the first variances. See, among others, Robert Fogelson, *Downtown: Its Rise and Fall* (New Haven: Yale University Press, 2001), 160–67.
17. John Logan and Harvey Molotch, *Urban Fortunes: The Political Economy of Place* (Berkeley: University of California Press, 1987), 19. Logan and Molotch draw much of their emphasis on the common interest of area residents from the "collective consumption" of goods in the private and public market put forth by Manuel Castells, *The Urban Question.* (Boston: MIT Press, 1979).
18. This work is not intended to be a history of Progressives and the movement in general. I focus in several chapters on the efforts of Progressive reformers and the types of knowledge they formed and promoted, concentrating mainly on urban Progressive groups like the Bureau of Municipal Research, who emphasized a Taylorized approach to the management of space, work, and leisure. I am more interested in how urban space creates the conditions for relations of power that allow these reform groups to

dominate public discourse and set many of the terms of urban restructuring. These reformers actively promoted the idea of "universal" citizenship based on the daily participation of urban residents in the political process, encouraging citizens to "demand efficiency" from city government. Their efforts were part of a larger trend in Progressivism after 1900 toward scientific approaches to reform and an emphasis on integrating the urban poor into consumer society. See William Allen, *Universal Training for Citizenship and Public Service* (New York: Macmillan, 1919); Frederick Cleveland, *Efficient Citizenship* (New York: Longman and Green, 1913); Simon Patten, *The New Basis of Civilization* (New York: Longman and Green, 1906).
19. Cleveland, *Efficient Citizenship*, 34.
20. On the expansion of the administrative capacity of the federal government, see Stephen Skowronek, *Building a New American State* (Cambridge: Cambridge University Press, 1982). On the restrictions on immigration, see Roger Daniels, *Guarding the Golden Door* (New York: Hill and Wang, 2004).
21. New York City developed a particular form of reform politics as a result of state election laws that allowed for fusion candidates, running on the party lines of two or more political parties. Successive mayoral administrations also forged working relationships with reform groups like the Bureau of Municipal Research, often to prove their reform credentials.
22. The concepts are drawn from the work of David Harvey and Henri Lefebvre, respectively.
23. David Harvey, *Justice, Nature and the Geography of Difference* (New York: Blackwell, 1996), 6. Though it is my intention to test several of Harvey's theoretical ideas, I do not consider this a work of historical-geographic materialism, for reasons that I hope will become clear. Whereas the basic question of the constitutive nature of produced space comes directly from Harvey, this work takes a slightly more mediated view of how space is actually produced, both in the particular and the abstract.
24. Ibid., 6.
25. David Harvey, *Spaces of Hope* (Berkeley: University of California, 2000), 75.
26. Ibid., 75.
27. Ibid., 430.
28. Eric Swyngedouw, "Neither Global nor Local: Glocalization and the Politics of Scale," in *Spaces of Globalization: Reasserting the Power of the Local*, ed. Kevin Cox (New York: The Guilford Press, 1997). As I will show in later chapters, shifts in economic networks of exchange at the larger scale greatly affected the Middle West Side long before the era of globalization, causing economic security to be a main source of tension in everyday life. The case of Higgins Carpet, which I deal with at length, illuminates this point. Also valuable on space and scale is Neil Smith, *Uneven Development: Nature, Capital, and the Production of Space* (London: Oxford University Press, 1990).
29. Logan and Molotch, *Urban Fortunes* 12.
30. Harvey, *Justice, Nature*. 1996, 44. A fundamental question is whether the process of uneven geographic development shifts from being structured by systems of domination described by Harvey and critical theorists becomes a process of hegemonic development described by Jean Baudrillard.

31. Ibid.
32. Harvey's reductive view of development, though the most comprehensive explication of capitalist scalar urban production of space, discounts the importance of factors that exist outside of the basic categories of labor, production, and consumption. Logan and Molotch attempt to account for these factors in combination with the classic Marxian categories, but are in themselves reductive in their economistic emphasis.
33. Soja, *Postmodern Geographies*, 132.
34. Henri Lefebvre, in *The Production of Space*, poses the question: "Does language—logically, existentially, genetically speaking—precede, accompany or follow social space? Is it a precondition of social space or merely a formulation of it?" He suggests that a case can be made for giving precedence in the construction or generation of language to "those activities which mark the earth," the altering of nature through human activity (16–17).
35. My approach to these layers of cognitive interpretation relies on critical-depth realism, as developed by Roy Bhasker and used in historical sociology by, among others, George Steinmetz. Roy Bhasker, *A Realist Theory of Science* (New York: Harvester Press, 1978); George Steinmetz, "Critical Realism and Historical Sociology," *Comparative Studies in Society and History* 40/1 (January 1998), 170–86.
36. I will argue in subsequent chapters that the publics that are in part produced by the changing spatial environment, such as the reform community, read the causal mechanisms of the spatial landscape differently.
37. Robert Beauregard, *Voices of Decline: The Postwar Fate of U.S. Cities*, 2nd ed. (New York: Routledge, 2003), xi.
38. Lefebvre, *The Production of Space*, 46. Emphasis in original.
39. Ibid., 413.
40. Kevin Lynch, *The Image of the City* (Cambridge, MA: MIT Press, 1960). Lynch's work explores the "look" of the city, and how the visual image governs movement.
41. Lefebvre, *The Production of Space*, 34.
42. Ibid., 38.
43. M. Christine Boyer, *Dreaming the Rational City; The Myth of American City Planning* (Cambridge, MA: MIT Press, 1983). In spite of the critique of some of Boyer's subtleties I consider her work to be one of the few attempts to perform a spatial analysis of the Progressives.
44. Michel Foucault, *Technologies of the Self: A Seminar with Michel Foucault* (Amherst: University of Massachusetts Press, 1988).
45. Ira Katznelson, *City Trenches: Urban Politics and the Patterning of Class in the United States* (Chicago: University of Chicago Press, 1981), 65–66.
46. Andrew Herod, "Workers, Space and Labor Geography," *International Labor and Working-Class History* 64 (Fall 2003): 112–38. See also Doreen Massey, *Spatial Divisions of Labor: Social Structures and the Geography of Production* (London: Blackwell Publishing, 1995).
47. My contention is that the inherent logic of both liberal democracy and capitalist accumulation, operating below the cognitive radar of collective actors, requires that homogeneity never be achieved in either spatial or temporal forms.
48. Ernst Bloch, "Nonsynchronism and Its Obligation to Dialectics," in *New German Critique* 11 (Spring 1977), originally published in 1932

in *Erbschaft dieser Zeit*. Bloch was of course writing about the temporal, and was one of many neo-Marxists who considered spatial analysis "reactionary."
49. Judith Shklar, *American Citizenship: The Quest for Inclusion* (Cambridge: Harvard University Press, 1991), 1–3.
50. For instance, universal male suffrage was the law, women were excluded from voting, and the right to take legal, collective action in the negotiation of wages and working conditions was in constant flux.
51. Rogers Smith, *Civic Ideals: Conflicting Visions of Citizenship in U.S. History* (New Haven: Yale University Press, 1997), and *Stories of Peoplehood: The Politics and Morals of Political Membership* (Cambridge: Cambridge University Press, 2003).
52. Smith, *Stories of Peoplehood*, 11–12.
53. Jonathan Boyarin, "Space, Time and the Politics of Memory," in *Remapping Memory: The Politics of TimeSpace*, ed. Jonathan Boyarin (Minneapolis: University of Minnesota Press, 1994).
54. *New York Times*, April 6, 1905. The *Times* favored stories about efforts at reforming the "lower" classes, which are, of course, embellished. They also published sensationalized accounts of crime in the Middle West Side, as well as "travel" pieces about slumming.
55. David Hammack, *Power and Society in Greater New York* (New York: Columbia University Press, 1985). The ubiquity of these images start with reformers like Jacob Riis, and now include institutions like the Lower East Side Tenement Museum.
56. The figure of the tenement slum dweller who rarely if ever leaves the immediate vicinity of home is a recurring theme of reform and journalistic accounts, and is usually represented as a woman. It was assumed that men had access to wider experiences, but women were strictly tied to the home.
57. The language of crisis took many forms in Progressive Era writing, from dire predictions regarding immigration to the squalid conditions of the city.
58. Contemporary work on "underserved" urban areas remains contained within a language of moral uplift, even if the work is careful to avoid the paternalism of Progressive Era reformers.
59. Robert Beauregard. *Voices of Decline: The Postwar Fate of U.S. Cities*, 2nd ed. (New York: Routledge, 2003). Beauregard examines the ways in which public and private investment is framed by the language of urban crisis.
60. On the history of urban land use, zoning, and scales, see Harvey, *The Urban Experience*.
61. The relationship between Progressive Era reform and the modernization of the United States is the theme of many works. On the development of the industrial economy, Samuel Hays, *The Response to Industrialism, 1885–1914* (Chicago: University of Chicago Press, 1957), is still a valuable overview. On the middle-class nature of Progressivism, see Michael McGerr, *A Fierce Discontent: The Rise and Fall of the Progressive Movement in America* (New York: Oxford University Press, 2003).
62. McGerr, *A Fierce Discontent*, 19.
63. Bruere, *New City Government*, 45.

64. The work of Jacob Riis is often regarded as a turning point in bringing images of the urban poor to the attention of the wider public. See Daniel Czitrom and Bonnie Yochelson, *Rediscovering Jacob Riis: Exposure Journalism and Photography in Turn-of- the-Century New York* (New York: New Press, 2008).
65. The link between space and mental and physical health came from a variety of professional fields, from public health workers to architects. Excellent examples can be found at http://www.fordham.edu/academics/colleges_graduate_s/undergraduate_colleg/fordham_college_at_l/special_programs/honors_program/hudsonfulton_celebra/homepage/progressive_movement/tenements_32232.asp.
66. Lawrence Veiller and Robert DeForest, *The Tenement House Problem* (New York: Macmillan and Sons, 1903).
67. Jacob Riis, *How the Other Half Lives* (New York: Dover Publications, 1971).
68. Real housing reform and effective enforcement of building code legislation begins after 1901. Zoning regulation that separated housing from industry starts in 1916. On housing reform, Richard Plunz, *A History of Housing in New York City: Dwelling Type and Social Change in the American Metropolis* (New York: Columbia University Press, 1990), 194–96. On zoning, see Max Page, *The Creative Destruction of Manhattan, 1900–1940* (Chicago: University of Chicago Press, 1999), 61–65.
69. Chris Otter, "Making Liberalism Durable: Vision and Civility in the Late Victorian City," *Social History* 27/1 (January 2002).
70. Ibid., 3.
71. DeForest and Veiller, 34–35.
72. I. M. Phelps Stokes, 12.
73. The Four Hundred was a list of New York's social elite compiled by journalists at the turn of the century. Hammack, 167.
74. Goldmark, *West Side Studies*, 64.
75. Ibid., 34.
76. Ibid., 69. Many of Goldmark's more lurid descriptions trail off in the language, leaving details to the prurient imagination of the reader.
77. Ibid., 66.
78. Ibid., 67.
79. Ruth True, *The Neglected Girl* (New York: Russell Sage Foundation, 1908).
80. David Kopel, http://www.davidkopel.org/CJ/Mags/InnerCityCrisis.htm. Greg Forster and Jay Greene, "Sex, Drugs, and Delinquency in Urban and Suburban Public Schools," Education Working Paper 4, Manhattan Institute for Policy Research.
81. Robert Wiebe, *The Search for Order, 1877–1920* (New York: Hill and Wang, 1967).
82. McGerr, 43.
83. Boyer, 29.
84. Roy Rosenzweig and Elizabeth Blackman, *The Park and the People: A History of Central Park* (Ithaca, NY: Cornell University Press, 1992).
85. *New York Times Supplement*, 1895.
86. Ibid.
87. Hammack, *Power and Society*, 145–46.

88. Anthony Jackson, *A Place Called Home: A History of Low-Cost Housing in Manhattan* (Cambridge, MA: MIT Press, 1976), 56–58.
89. Hammack, *Power and Society,* 132–34. The impression that Tammany dominated New York Democratic Party politics has been challenged by many historians, including Hammack, who points out that "machines" for patronage and votes were regional in the city.
90. The popularity of commissions, nonpartisan governing institutions, extended all the way to city government itself. Starting in Galveston, Texas, after a disastrous hurricane in 1902, commission government, with rotating mayors, spread to over a hundred cities in the United States by 1915.
91. Veiller, *Tenement House Problem,* 2.
92. Ibid., 8.
93. The act securing construction of the park was written by Thomas McManus, the patriarch of the Hell's Kitchen political family that established the McManus Club in 1898.
94. Plunz, *History of Housing,* 23.
95. *Real Estate Record* (Journal). No. 65, May 26, 1900.
96. Plunz, *History of Housing,* 96.
97. New York State Court of Appeals.
98. Burrows and Wallace, *Gotham,* 456.
99. Jackson, *Place Called Home,* 54.
100. "City Sanitary Progress Traced by Charles Wingate," *New York Times,* September 8, 1897.
101. Plunz, *History of Housing,* 213–14.
102. Cartwright, *Middle West Side,* 50–54.
103. Ibid., 61.
104. Lynch, *Image of the City,* 90–96.
105. Anthony, *Women Who Earn,* 34.
106. Ibid., 65. Anthony's survey showed that the majority of married women with children worked in domestic or cleaning work, and the majority of unmarried women worked in manufacturing.
107. Paddy's was a collection of licensed and unlicensed pushcart vendors. The market was eventually shut down in 1940.
108. Though urban historians have written much about the supposed "solidarity" of working-class neighborhoods, recent research is showing just how fragmented and transient these areas could be. See, for example, Alaine Faure, "Local Life in Working Class Paris at the End of the Nineteenth Century," *Journal of Urban History* 32/5 (July 2006): 761–72.
109. See the review essay by Vyjayanthi Rao, "Slum as Theory: The South Asian City and Globalization," *International Journal of Urban and Regional Research* 30/1 (March 2006): 225–32; and also Partha Chaterjee, *The Politics of the Governed* (New York: Columbia University Press, 2004); and Arjun Appadurai, "Spectral Housing and Urban Cleansing: Notes on Millennial Mumbai," *Public Culture* 12/3: 2000, 627–51. Both Chaterjee and Appadurai suggest that the performative nature of illicit activity in underserved areas is a condition of modernity and a strategy for making demands upon the modern state. I am drawing an asymmetrical parallel with conditions in urban areas in the U.S. Progressive Era.

110. All quotes from Middle West Side residents residing in the area between 1900 and 1915, from Jeff Kissellhoff, *You Must Remember This* (Baltimore: Johns Hopkins University Press, 1989), 576–79.
111. *New York Times*, July 5, 1911.
112. Doreen Massey *Spatial Divisions of Labor: Social Structure and the Geography of Production* (London: Macmillan, 1984).
113. Burrows, *Gotham*, 345–47.
114. Richardson, 113.
115. "Vegetables the Poor Eat," *New York Times*, March 15, 1903.
116. For a discussion of evolving rights and public space, see Joseph Varga, "For Speaking Jewish in a Jewish Neighborhood: Civil Rights and Community-Police Relations During the Postwar Red Scare, 1919–1922," in *Uniform Behavior: Police Localism and National Politics*, ed. Stacy McGoldrick and Andrea McArdle (New York: Palgrave Macmillan, 2006), 60–65.
117. Marilynn Johnson, *Street Justice: A History of Police Violence in New York City* (Boston: Beacon Press, 2003), 2.
118. James Lardner and Thomas Reppetto, *NYPD: A City and Its Police* (New York: Henry Holt, 2000), 40.
119. Numerous instances of police graft involving legitimate businesses that were regulated by the city exist. For an interesting account, see "Investigating New York's License Bureau," *Charities and the Commons* (May 1908): 143–47.
120. Johnson, 87–90. Roosevelt's reforms were concerned with breaking the relationship between the police and his political opponents, and with "professionalizing" the force through better recruitment, training, new uniforms, codified procedures, and the use of standardized weapons and tactics.
121. The committee was named for Clarence Lexow, an upstate state senator. The forming of the committee was brought about through pressure from reformers like Parkhurst, but was also part of a long-standing and ongoing struggle between the city and the state legislature over control of the metropolitan region. As Johnson points out, the committee was not concerned, in its formation, with issues of violence or brutality but with impunity in the form of lawbreaking through graft, bribery, and corruption. The testimony of New Yorkers from all social classes regarding violent treatment at the hands of NYPD officers did bring the issue into the public sphere, though issues of coordinated violence against labor groups and brutal treatment of working-class citizens received less attention than other, politically motivated issues.
122. "Report of the Special Committee Appointed to Investigate the Police Department of the City of New York," January 18, 1895, in *New York City Police Corruption Investigation Commissions, 1894–1994*, vol. 1, ed. Gabriel J. Chin (Buffalo, NY: W.S. Hein, 1997).
123. For an interesting discussion of spatialized policing and its long connection to the liberal state and evolving notions of legality and rights, see Don Mitchell, "The Liberalization of Free Speech: Or, How Protest in Public Space is Silenced, " *Agora* 4 (2003).
124. Gilfoyle, 200–202.
125. From the period of the Tweed scandals (1874), through the reform era, Tammany-connected administrations maintained a fragile electoral

coalition. It was held together by the clubhouses in each assembly district that handled patronage, populist appeals to workers and organized labor, and cooperation with certain wealthy New Yorkers, particularly those invested in maintaining private market control over utilities, transport, and housing. See Sven Beckert, *The Monied Metropolis*, and Hammack, *Power and Society in New York City*.
126. Lardner and Reppetto, *NYPD*, 21.
127. Lardner and Reppetto, *NYPD*, 100–102.
128. For the period of the Lexow Committee and subsequent push for police reform, the *New York Press, New York Tribune*, and *New York Times* were anti-Tammany papers, whereas the *New York Sun, Herald*, and *World* were more populist and leaned toward the local Democratic Party. Of course, these loyalties were fluid and shifted over time. For the political divisions and syntactical style of the papers, see Christopher Wilson, *The Labor of Words: Literary Professionalism in the Progressive Era* (Athens: University of Georgia Press, 1985), and David Ray Papke, *Framing the Criminal: Crime, Cultural Work and the Loss of Critical Perspective* (Hamden, CT: Archon Books, 1987).
129. Among many examples, see "Seeing Manhattan: Saturday Night in Hell's Kitchen," *New York Times*, July 16, 1905.
130. George Chauncey, *Gay New York: Gender, Urban Culture, and the Making of the Gay Male World, 1890–1940* (New York: Basic Books, 1994), 139–49. Chauncey's study is an important contribution to a spatialized understanding of the modern industrial city and how space was "claimed" and contested by different groups.
131. All statistics from *Yearly Report City of New York Magistrates Court First Division* compiled by the Police Clerk and reported to the Board of City Magistrates, 1908.
132. Chauncey, *Gay New York*, 172.
133. James Richardson. *The New York Police: Colonial Times to 1901* (New York: Oxford University Press, 1970). But as Richardson points out, the instituting of civil service reform in police hiring was a highly contested process, with the police board maintaining some control over the process after 1884, the year civil service reform was adopted. The board would be allowed to choose from a list of names of those who had passed the civil service exam, allowing for influence and graft in the hiring process.
134. Goldmark, *West Side Studies*, 13.
135. Ibid., 14.
136. "A Policeman Who Captured More than He Wanted," *New York Times*, September 19, 1902.
137. Goldmark, *West Side Studies*, 10–13.
138. Charles B. Barnes, *The Longshoremen* (New York: Survey Associates, 1915). The earliest dock strike took place in 1836 over wages and hours. Major strikes also occurred in 1853, 1874, 1887 (organized by the Knights of Labor), and 1896. These major actions included dockworkers from the West Side, East Side, Brooklyn, and New Jersey. Numerous smaller actions took place on a regular basis.
139. Richardson, 204. http://www.pps.org/.
140. *Annual Report of the New York City Police Department, 1895–1910*. The report on officers' places of residence showed that the majority of

patrolmen who lived on the West Side tended to congregate away from the river, toward Seventh and Eighth Avenues. None lived west of Ninth Avenue.

141. *People of NYC vs. Arthur Harris,* New York City Magistrate Court, October 29, 1900. Harris testified that he felt his life was in danger from the clubbing, and did not realize Thorpe was a police officer. He also testified that Thorpe used a series of racial epithets while beating him.

142. "War in Hell's Kitchen," *New York Times,* August 4, 1898. The *Times* story is one of numerous sources from the period that claim that landlords actively sought out African American renters, who were considered more reliable tenants. For an analysis from the period, see Robert E. Park, "Negro Home Life and Standards of Living," *Annals of the American Academy of Political and Social Science* 32 (1904): 147–63.

143. "The Story of the Riot," *New York Daily Tribune,* August 16, 17, 18, 1900; "The Story of the Riot: Persecution of Negroes by Roughs and Police," *New York Times,* August 16, 1900. See also Gilbert Osofsky, "Race Riot 1900: A Study of Ethnic Violence," *Journal of Negro Education* 32/ 1: 16–27.

144. Chief Devery was also responding to severe criticism of his actions during white riots in 1898. On the police riot, see Richard O'Connor, *Hell's Kitchen: The Roaring Days of New York's Wild West Side* (New York: Lippincott, 1955), 145–57, which draws its narrative from newspaper reports and the writings of reformer Frank Moss concerning the incidents of police brutality.

145. David Roediger, *Working toward Whiteness: How America's Immigrants Became White* (New York: Basic Books, 2005), 62.

146. Ibid., 63.

147. The number of people of non-European origin, such as Asians or those from the Middle East, was negligible in the area under study. Small pockets of Asians lived in Hell's Kitchen, along with some North Africans.

148. *Department of Parks Annual Report,* 1902. http://www.nycgovparks.org/news/reports/archive#ar

149. The 1902 *Annual Report of the Department of Parks* provides not only a description of the changes wrought by DeWitt Clinton's near-completion, but ambitious plans for future park construction, most of which were never carried out.

150. The annexation of Hawaii, and the potential annexations of Cuba and the Philippines in the 1890s had sparked serious debates about expansionism, empire, and the potential assimilation of what Warren Harding had referred to as "our little brown brothers." Isolationists in this debate doubted whether Cubans or Filipinos could ever be "self-governing citizens," in language very similar to that deployed in contests over the "assimilability" of new immigrants.

151. The traditional starting point to explore American anti-urbanism is Thomas Jefferson's insistence that democracy and the city were absolutely incompatible, a position that Jefferson softened later in his life. The tradition includes writers from Thoreau to Mumford and John Dewey. A dated but still relevant review of intellectual anti-urbanism in the United States is Morton and Lucia White, "The American Intellectual versus the American City," *Daedalus* (Winter 1961): 166–79.

152. Frederick C. Howe, *The City: The Hope of Democracy* (New York: Charles Scribner and Son, 1905). Howe is influential in acknowledging that real, existing cities are the key nodal points for the development of a modern economy, in opposition to the anti-urbanism of people like Ebenezer Howard. Howe acknowledges Jefferson's points regarding the difficulties of combining large population clusters with extreme disparities of wealth with a truly functioning democracy. His solution, among other suggestions, is the single tax on land, drawn mainly from the populist Henry George.
153. Daniel T. Rodgers, *Atlantic Crossings: Social Politics in a Progressive Age* (Cambridge, MA: Harvard University Press, 1998), 112.
154. Lawrence Veiller, *A Plan for Tenements in Connection with a Municipal Park* (New York: Scribner and Sons, 1903).
155. See for example the website for the Project for Public Space, http://www.pps.org/, with its emphasis on place making and community.
156. For an enlightening discussion of the methodological issues involved in the space-place dichotomy, see Andrew Sayer, "Behind the Locality Debate: Deconstructing Geography's Dualisms," *Environment and Planning* 23 (1991): 283–308.
157. Steven Harrison and Paul Dourish, *Re-Placing Space: The Role of Space and Place in Collaborative Systems*. Proceedings of the CSCW ACM, 1996.
158. J. Nicholas Entrikin, *The Betweenness of Place: Toward a Geography of Modernity* (Baltimore: Johns Hopkins University Press, 1991), 5–7.
159. Yi Fu Tuan, *Space and Place* (Minneapolis: University of Minnesota Press, 1977), 6.
160. Quoted in Entriken, *The Betweenness of Place*, 68.
161. John Loganand Harvey Molotch, *Urban Fortunes: The Political Economy of Place* (Berkeley: University of California Press, 1987).
162. Doreen Massey, *Spatial Divisions of Labor: Social Structure and the Geography of Production* (New York: Routledge, 1995).
163. For an excellent analysis of these distinctions, see Entrikin, *The Betweenness of Place*.
164. Michel Foucault, "Of Other Spaces," http://foucault.info/documents/heteroTopia/foucault.heteroTopia.en.html.
165. Isaac Newton Phelps Stokes, *Iconography of Manhattan Island* (New York: Robert H. Dodd. The iconography was published in multiple volumes between 1915 and 1928.
166. Patten's analysis calls for a new outlook that scorned the assumption of scarcity and embraced abundance and ways to equitably distribute this abundant surplus. Simon Patten, *The New Basis of Civilization* (1905; repr., Cambridge, MA: Bellknap, 1968).
167. The question of how to define and classify the amorphous group "Progressives" during this period is a historian's cottage industry and would require a complete bibliography of Progressive Era history. Some recent histories of the era have been more specifically spatial, such as Robin Bachin's work on Chicago and Jonathan Hepp's work on Philadelphia, cited below.
168. Benedict Anderson, *Imagined Communities: Reflections on the Origin and Spread of Nationalism* (London: Verso Press, 1991) 34–36.

169. Appadurai, Arjun, *Disjuncture and Difference in the Global Economy*, http://www.intcul.tohoku.ac.jp/~holden/MediatedSociety/Readings/2003_04/Appadurai.html.
170. Frederick C. Howe, "The Remaking of the American City," *Harper's Monthly Magazine* 127 (July 1913): 186.
171. Benjamin Marsh, "City Planning in Justice to the Working Population," *Charities and the Commons* 19 (February 1908): 1514.
172. John Henry Hepp, *The Middle-Class City: Transforming Space and Time in Philadelphia, 1876–1926* (Philadelphia: University of Pennsylvania Press, 2003), 1–2; Robin F Bachin, *Building the South Side: Urban Space and Civic Culture in Chicago, 1890–1914* (Chicago: University of Chicago Press, 2004), 6.
173. Keller Easterling, *Enduring Innocence: Global Architecture and Its Political Masquerades* (Cambridge, MA: MIT Press, 2004), 2.
174. Quote and image from Ebenezer Howard, *To-Morrow: A Peaceful Path to Real Reform,* repr. as *Garden Cities of Tomorrow* (London: Faber and Faber, 1902).
175. Ernest Flagg, "The Plan of New York and How to Improve It," *Scribner's Magazine* 36 (August 1904): 253.
176. Ibid., 256.
177. Rodgers, *Atlantic Crossings,* 3.
178. Frank Koester, "American City Planning," *American Architect* 102 (October 23, 1912).
179. Flavel Shurtleff, "The English Town Planning Act of 1909," in *Proceedings of the Second National Conference on City Planning* (Boston: National Conference on City Planning, 1910), 179–80.
180. Ibid., 184.
181. Frederick Law Olmsted and James Croes, *Preliminary Report of the Landscape Architect and the Civil and Topographical Engineer* (New York: Department of Public Parks, 1876), Document 72; John C. Olmsted, "The Relation of the City Engineer to Public Parks," *Journal of the Association of Engineering Societies* 13 (October 1894): 594–95.
182. Benjamin Marsh, "Economic Aspects of City Planning," *Proceedings of the Municipal Engineers of New York,* Paper no. 57, 1910.
183. Andrew Jonas, "The Scale Politics of Spatiality," *Environment and Planning D: Society and Space* 12 (1994): 262.
184. Benjamin Marsh, "The Causes of Congestion," *Proceedings of the Second National Conference on City Planning* (Boston: National Conference on City Planning, 1907), 35–39.
185. Charles Lamb, "The City Plan," *The Craftsman* 3/13 (April 1904).
186. Flagg, "The Plan of New York and How to Improve It," 254.
187. Scale is used here to represent the different connective tissues linking Hell's Kitchen residents to other forms of social organization, such as individual households and regional economic processes. The use of scale in analysis of human geography has recently come under question. For a good summation of the debate see Marston, Jones, and Woodward, "Human Geography without Scale," *Journal of the Institute of British Geographers* 30 (2005).
188. Anthony, *Women Who Earn,* 27–28.
189. Entrikin, *The Betweenness of Place,* 126–27.

Notes to pages 142–155

190. The literature on ethnic solidarity is, of course, extensive, particularly for neighborhoods housing newly arrived and first-generation immigrant groups.
191. For instance, Halcyon Hall, at Third Avenue and 60th Street, seemed a popular location.
192. *Catholic Register,* 1910.
193. John T. Ridge, "Irish County Societies in New York, 1880–1914," in *The New York Irish,* ed. Ronald Boyer and Timothy Meagher (Baltimore: Johns Hopkins University Press, 1996), 301–20.
194. "Celtic Park Hosts Local Games and Families," *Gaelic American,* April 12, 1902.
195. St. Raphael's is an excellent example of the shifting ethnic boundaries of the area. The parish relocated from 40th and Eighth to its present location, a stunning neo-Gothic structure near the Lincoln Tunnel entrance on Eleventh Avenue. As Irish communicants left the area, the church was taken over by Italian Franciscans in the 1930s and eventually became St. Cyril and Methodius, a Croatian church whose original home was a converted tenement on West 50th Street.
196. Marion Casey, "Irish Americans on the Move in New York City," in Bayor and Meagher, *The New York Irish,* 395–418.
197. Colleen, McDannell, "Going to the Ladies Fair: Irish Catholics in New York City" in Bayer and Meagher, *The New York Irish,* 234–51.
198. Department of Docks and Ferries, *Annual Report,* 1894–1905.
199. Tom McConnon, *Angels in Hell's Kitchen* (New York: Doubleday, 1959). McConnon's book is an over-romanticized reminiscence written fifty years after the fact, but contains valuable insight into the daily lives of local families.
200. Ibid., 34–36.
201. Ibid., 56–67.
202. Department of Parks, *Report for the Year 1902.*
203. Exhibits would become the rage among the reforming classes through the 1900s.
204. The Citizens Union was one of many elite-based civic organizations that recognized the benefits of reforming space in underserved areas.
205. "These City Children Will Learn Farming," *New York Times,* July 27, 1902.
206. Department of Parks, *Report for the Year 1903.*
207. Goldmark, *West Side Studies,* 5.
208. Mary Brendle, *Clinton/Hell's Kitchen and Its Women* (New York: Harcourt Press, 1997).
209. *New York Times,* December 12, 1913.
210. For a detailed history of changes in housing styles, laws and regulations, see Oliver Plunz, *A History of Housing in New York City* (New York: Columbia University Press, 1990).
211. Ibid., 46–48.
212. "The Miseries of Poverty," *New York Times,* December 15, 1884.
213. Elsa Herzfeld, *Family Monographs: The History of 24 Families Living in New York's Middle West Side* (New York: Kempster Printing, 1905), 34–35.
214. "New Model Homes for West Side," *New York Times,* March 13, 1913.

215. The charge that working-class families rented few model apartment units was a consistent press item from 1890 to 1915.
216. "Model Homes Report," *New York Times*, June 2, 1918.
217. "Model Tenements Not a Failure," *New York Times*, January 19, 1913.
218. "Men Crowded Out as Tenement Heads," *New York Times*, May 21, 1909.
219. Anthony, "The World of the West Side Mothers," in *Women Who Must Earn*, 8–23.
220. "Sacred to the New Woman," *New York Times*, April 10, 1998.
221. New York City Police Census, 1900 and 1910.
222. Ibid.
223. Esther Packard, *Consumer League Report on Housing for Working Women*, 1914.
224. Anthony, "The World of the West Side Mothers," 71–72.
225. "Socialized Flats for Working Girls," *New York Times*, February 13, 1916.
226. Packard, *Consumer League Report on Housing for Working Women*, 21.
227. Herzfeld, *Family Monographs*, 56–59.
228. Ibid., 34–36, 39–41.
229. Ibid., 50–51.
230. "Devery Entertains at Big Barbeque," *New York Times*, September 2, 1902.
231. One can trace this particular brand of anti-urbanism from Ebenezer Howard through Lewis Mumford and beyond.
232. The persistent discourse of the urban problem, from the "super-predator" scare of 1990s "crack babies" to today's overblown rhetoric regarding immigrant gangs and "home invasion" are testament to the lasting effects, as are contemporary quests of place making. See Adolph Reed, *Stirrings in the Jug: Black Politics in the Post-Segregation Era* (Minneapolis: University of Minnesota Press, 1999).
233. Warren Magnusson, "The City of God and the Global City," *C-Theory*, October 5, 2006, http://www.etheory.net/articles.aspx?id=520.
234. "Higgins Carpet Works to Go," *New York Times*, October 8, 1900.
235. *American Carpet and Upholstery Journal* 25 (November 1900); "Heard about Town," *New York Times*, October 14 and 19, 1900.
236. The only detailed account of Higgins Carpet remains several chapters devoted to the carpet business and global and local economies in John Ewing and Nancy Norton, *Broadlooms and Businessmen* (Cambridge, MA: Harvard University Press, 1955).
237. Harvey, *The Urban Experience*, 33.
238. Jefferson Cowie, *Capital Moves: RCA's 70-Year Quest for Cheap Labor* (New York: New Press, 2001).
239. Harvey, *The Urban Experience*, 101–4.
240. Andrew Herod, "Workers, Space and Labor Geography," *International Labor and Working-Class History* 64 (Fall 2003): 112–38.
241. Massey, *The Spatial Division of Labor*.
242. Burrows and Wallace, 455–58.
243. New York Board of Aldermen, 1900 Crosby Codes.
244. Anthony, *Women Who Earn*, 45.
245. True, 67.
246. Ibid., 69–70.
247. Lawrence B. Glickman, *A Living Wage: American Workers and the Making*

of Consumer Society (Ithaca, NY: Cornell University Press, 1997).
248. As Mary Brendle's work on Hell's Kitchen (Clinton) demonstrates, many still live there. Several bakeries, meat shops, and one ice business have been operating in Hell's Kitchen under the ownership of one resident family, since the late nineteenth century.
249. Katherine Anthony, *Mothers Who Must Earn* (New York: Russell Sage Foundation, 1914).
250. Alice Kessler-Harris, *Out to Work: A History of Wage Earning Women in the United States* (New York: Oxford University Press, 1982), 119.
251. Ibid., 120.
252. Ibid., 109.
253. Anthony, *Mothers Who Must Earn*, 88.
254. Ibid., 70.
255. Ibid., 27.
256. Ibid., 78. All names in the Anthony report are fictional.
257. Ibid.
258. Ibid., 85–86.
259. Most investigations of the laundry industry found numerous abuses of the lax overtime laws then in force.
260. Anthony, *Mothers Who Must Earn*, 76.
261. Average wages in 1911 in Middle West Side laundries ranged from $4 per week for the least skilled jobs, through $6.50 to as much as $12 per week for some supervision. Ibid., 118.
262. Sue Clark and Edith Wyatt, *Making Both Ends Meet* (New York: Macmillan, 1911). Clark and Wyatt's work grew out of a study commissioned by the Consumers League of New York City. Conditions in laundries were also extensively studied by the U.S. Department of Labor, which issued a report, *Employment of Women in Laundries*, the same year.
263. Anthony, *Mothers Who Must Earn*, 57.
264. Iris Marion Young, *Justice and the Politics of Difference* (Princeton: Princeton University Press, 1990), 174.
265. David Harvey, "Class Relations and Social Justice," in *Place and the Politics of Identity*, ed. Michael Keith and Steven Pile (London: Routledge, 1993), 56.
266. Anthony, *Mothers Who Must Earn*, 83.
267. Isabel Dyck, "Feminist Geography, the Everyday, and Local-Global Relations: Hidden Spaces of Place-Making," *Canadian Geographer* 49/3: 233–43.
268. Barnes, *Longshoremen*, 3.
269. Calvin Tompkins, "The Desirability of Comprehensive Municipal Planning in Advance of Development.,"*Proceedings of the Municipal Engineers of the City of New York for 1905* 19/3.
270. Ibid., 6.
271. An excellent summary of the history of the New York City docks and piers can be found in the collection of essays *The New York Waterfront: Evolution and Building Culture of the Port and Harbor*, ed. Kevin Bone (New York: Monticelli Press, 1997).
272. Hendrik Hartog, *Public Property and Private Power: The Corporation of the City of New York in American Law, 1730–1870* (Ithaca, NY: Cornell University Press, 1983), 119.
273. Ibid, 120–123.

274. Cartwright, Middle *West Side*, 27–29.
275. New York City Department of Docks and Ferries, *Annual Report*, 1896.
276. On Plunkitt, see Avinash Dixit and John Londregan, "The Determinants of Success of Special Interests in Redistributive Politics," *Journal of Politics* 58/4 (November 1996): 1132–55.
277. New York City Department of Docks and Ferries, *Report on the Scope and Limits of Expropriation*, April 1912.
278. The construction of the recreation pier will be discussed in more detail in chapter 6.
279. New York City Department of Docks and Ferries, *Annual Report*, 1907.
280. The telegraph was installed for the West Side piers in 1896. Barnes, 27.
281. Ibid., viii.
282. Ibid., iv–v.
283. Ibid., 4.
284. Ibid., 11–12.
285. Ibid., 23–24.
286. Ibid., 47–48.
287. Ibid., 113.
288. Ibid., 66.
289. See, for instance, David Montgomery's *Worker's Control in America: Studies in the History of Work, Technology, and Labor Struggles* (Cambridge: Cambridge University Press, 1980).
290. Barnes, *Longshoremen*, 104–11.
291. Ewing and Norton, 64–73.
292. Ibid., 88.
293. "Strikers to Return to Work," *New York Times*, January 29, 1899.
294. Ibid.
295. "The Higgins Strike," *New York Times*, February 8, 1899.
296. Bruno Latour, "On Actor Network Theory: A Few Clarifications," http://www.nettime.org/List-Archives/nettime-1-9801/.
297. Noel Castree, "Labor Geography: A Work in Progress," *International Journal of Urban and Regional Research* 31/4 (December 2007): 853–62.
298. Giovanni Arrighi, *The Long Twentieth Century*, and Immanuel Wallerstein.
299. Don Kalb, "Class (in Place), without Capitalism (in Space)?," *International Labor and Working-Class History* 57 (Spring 2000) 36. Even here, Kalb hints at a wider view beyond labor geography by suggesting that capitalism is "world-embedded."
300. Herod, *Workers, Space, and Labor Geography*, 114.
301. Katznelson, *City Trenches*, 45–49.
302. Sallie Marston, John Paul Jones III, and Keith Woodward, "Human Geography without Scale," *Transactions of the Institute of British Geographers* 30 (2005), 416–32.
303. Ibid., 428–29.
304. Letter to the Editor, *New York Times*, May 18, 1911. Between 1890 and 1920, the Middle West Side was consistently cited by city residents in newspaper letters as either an example of city corruption and lack of services or as the blank slate on which to work reform.
305. "Not Hell's Kitchen," *New York Times*, Letter to Editor, May 16, 1900.
306. A review of the ethnic and labor press available to Middle West Side residents after 1900 reveals an interesting tension between a defense of

the local area as belonging to the working residents and encouragement of those same residents to own their own home in the expanding "suburbs" in the outer boroughs and Long Island. Examples of both are found in the *Irish American, Gaelic American,* and several trade union journals.
307. Paula Bello, "The Shifting Global Landscapes of Things: An Interview with Arjun Appadurai," *Design and Culture* 2/1 (March 2010): 65–78.
308. Mary Simkhovitch, "The Function of a Social Settlement," *Charities,* May 1902.
309. This discussion of networks is largely drawn from the work of Bruno Latour's actor/network theory. See Bruno Latour, *Reassembling the Social: An Introduction to Actor/Network Theory* (Oxford: Oxford University Press, 1991).
310. There were numerous overviews of immigration during this period. I have mainly relied on Daniels, *Guarding the Golden Door*; and Desmond King, *Making Americans* (Cambridge, MA: Harvard University Press, 2000). Following the work of Jose Casanova, Aristide Zolberg, and others, I take the view that immigrant groups are incorporated into citizenship structures rather than assimilated.
311. Thomas Kessner, *The Golden Door: Italian and Jewish Immigrant Mobility in New York City, 1880–1915* (New York: Oxford University Press, 1977).
312. For an overview of this process, see John Higham, *Strangers in the Land: Patterns of American Nativism, 1860–1925,* 6th ed. (New Brunswick, NJ: Rutgers University Press, 2004).
313. The Russell Sage West Side Studies Series contains numerous examples of the tension between West Side residents and social reformers.
314. For a review of New York's role in the developing science of urban reform, see John Recchiuti, *Civic Engagement: Social Science and Progressive-Era Reform in New York City* (Philadelphia: University of Pennsylvania Press, 2007).
315. See Edwin Burrows and Michael Wallace, *Gotham: A History of New York City to 1898* (New York: Oxford University Press, 2000).
316. Clifton Hood, *722 Miles: The Building of the Subways and How They Transformed New York* (Baltimore: Johns Hopkins University Press, 1995).
317. Recent work on suburban working-class housing and its relation to industries moving from crowded cities has suggested that transport opportunities were not the universal driving cause behind the phenomenon. But Richard Harris makes the case that the subway and train system made New York City particularly susceptible to shifts in urban populations moving to the outer boroughs. Richard Harris, "Suburbanization and the Employment Linkage," in *Manufacturing Suburbs: Building Work and Home on the Metropolitan Fringe,* ed. Robert Lewis (Philadelphia: University of Pennsylvania Press, 2004), 221–36.
318. Kathy Peiss, *Cheap Amusements: Working Women and Leisure in Turn-of-the-Century New York* (Philadelphia: Temple University Press, 1986).
319. True, 72–73.
320. John F. Kasson, *Amusing the Millions: Coney Island at the Turn of the Century* (New York: Hill and Wang, 1978).
321. *New York Times,* May 19, 1905.
322. McConnon, *Angels in Hell's Kitchen,* 54–56.

323. A derisive term of the period for reformers, meaning goodie-goodie, and signifying naïveté and lack of understanding of how the city actually worked at the local level.
324. Recchiuti, *Civic Engagement*, 45.
325. "Motorists Now Safe In Hell's Kitchen," *New York Times*, August 19, 1908.
326. Thomas O'Donnell, "How the Irish Became Urban: The Irish Experience in Large American Cities," *Journal of Urban History* 25 (January 1998): 271–86.
327. Commerce and Industry Association, *Disposal of the West Side Tracks* (New York: Merchants Association of New York, 1908).
328. Much of this history of the Eleventh Avenue track is drawn from Calvin Tompkins's report to the New York City Department of Docks and Ferries, January 26, 1911.
329. *Report to the Board of Estimate and Apportion on the Removal of Eleventh Avenue Tracks*, December 1910, City of New York Hall of Records.
330. Tompkins, 1911.
331. "West Side Meeting Cheers Senator Saxe," *New York Times*, March 2, 1906.
332. "Long Fight Likely on Track Removal," *New York Times*, March 3, 1907.
333. Mayor William J. Gaynor, *Report to the New York Board of Estimate*, July 1910.
334. "Children Parade against Death Avenue," *New York Times*, October 25, 1908.
335. *New York Times*, October 29, 1908.
336. *New York Times*, May 11, 1911.
337. "A New Settlement House," *New York Times*, October 25, 1900.
338. Mina Carson, *Settlement Folk: Social Thought and the American Settlement Movement, 1885–1930* (Chicago: University of Chicago Press, 1990).
339. "Hartley House Receives Support," *New York Times*, December 10, 1896.
340. True, *Neglected Girl*, 6–7.
341. "Rockefeller Drops Settlement House," *New York Times*, January 15, 1910. An attempt by Senator McManus to obtain funds from the state for reopening the house as a nursery in 1911 had failed to pass.
342. New York City Department of Docks and Ferries, Annual Reports pf 1900, 1901, 1905. The issue of whether entertainment and refreshments should be franchised by the city to for-profit businesses or provided by the city at nonprofit rates was an issue of contention. For-profit franchising businesses won the dispute, and were often accused of patronage and of price gouging.
343. The battles between the city and pier owners were long-standing and seemingly endless. The city had, by 1907, seized four piers along the Middle West Side, two from Knickerbocker Ice, utilizing one, at 43rd Street, as a health center for checkups. In DoDF reports, dock masters (there were two for Manhattan), fought an ongoing struggle with the city to attain budget allocations for necessary repairs and dredging.
344. In 1896, Dock Master James Wheeler had requested a permanent police presence for the piers between 40th and 48th Streets.
345. Edwin Tracy, "New York's Recreation Piers Well Worth Their Costs," *New York Times,* Supplement, June 25, 1905.

346. "Says Hell's Kitchen Will Reform Itself," *New York Times*, March 20, 1912.
347. "Boys Demand Pier's Opening," *New York Times*, March 22, 1912.
348. *New York Times*, March 23, 1912.
349. An exhaustive search of New York City dailies from the 1880s to 1920 reveals very little public protest, at least as covered by the papers, before 1900. Most coverage concerned acts of criminality or group violence and riots before 1900. While those stories continued, incidents of local citizens acting in their own interest through public protest, including local merchants demanding more police protection (1908), increase after 1900.
350. *New York Times*, May 9, 1908.
351. Lefebvre refers to this as the "code" of any built city, the architectural logic of a specific mode of production.
352. The concept of community I am critiquing here is the commonly understood way of describing urban agglomerations of population that assume some long-standing, static ties between human space users based on ethnicity, income, consumption, or other unifying category. Myriam Puallac's historical work on Chicago's South Side black population is an excellent example of this type of useful yet limiting urban study. See Pauillac, "African-American Families, Urban Space, and the Meaning of Community Life," *Dialectical Anthropology* 26 (2001): 273–83.
353. Yi Fu Tuan, 199. Tuan's definition, "place is security," would seem to indicate that place can never exist under shifting patterns of spatial practice in a modern economy. A recent example of this use of place is John Friedman's "Reflections on Place and Place-Making in the Cities of China," *International Journal of Urban and Regional Research* 31/2 (June 2007): 257–79. Freidman suggests that under his definition place only exists when abstract space is "humanized," barring empty buildings and vacant lots the use of the term.
354. The legal right to vote was extended to all adults with the passage of the Nineteenth Amendment.
355. Of course, this list of rights is subject to interpretation and the evolving conceptions of rights worked out in the court system. I merely indicate that no distinction is made between the urban poor and other citizens in the official sense.
356. Harvey, *Justice, Nature,* 1996, 6–7.
357. For an excellent example of how perception affects development, see Robert Beauregard, *Voices of Decline: The Post-War Fate of U.S. Cities* (New York: Routledge, 2003).
358. Barnes, *Longshoremen,* 4.
359. Harvey, *Justice, Nature,* 1996, 123–25.
360. Wolfgang Shivelbush, *The Railroad Journey: The Industrialization and Perception of Time and Space* (Berkeley: University of California Press, 1987).
361. Barbara Adams and Adeola Enigkbo, "Critical Movements and Urban Space," unpublished ms., 2008.
362. Emily Talen, "Beyond the Front Porch: Regionalist Ideals in the New Urbanist Movement," *Journal of Planning History* 7/1 (February 2008): 20–47.
363. Soja, *Post-Modern Geographies,* 1989, 30.

364. Soja's work has been critiqued from many directions, including not being very postmodern, but his influence is undeniable. He makes a defense of his position and subsequent work in *Progress in Human Geography* 30/6 (2006): 812–20.
365. Jacobs was concerned with urban space long before Soja's call for a spatial turn. I refer here to the many urbanists influenced by her work, and her own contemporary work.
366. Revanchism and gentrification are the project of, for example, Neil Smith and his students. Neil Smith, "New Globalism, New Urbanism: Gentrification as Global Urban Strategy," *Antipode* 6/2 (2002).
367. Based in part on the theories developed by Marston, Jones, and Woodward, "Human Geography without Scale." They draw from post-structural materialist theories that include non-human actants as affective forces. See T. Schatzki, *The Site of the Social: A Philosophical Account of the Constitution of Social Life* (State College: Pennsylvania State University Press, 2002); Manuel Delanda, *A Thousand Years of Non-Linear History* (2005).
368. Lefebvre, *The Production of Space*. Emphasis in original.
369. Ibid., 36.

Index

Adler, Felix, 57–58, 73, 124, 138, 153, 154
Anderson, Benedict, 128
Anthony, Katherine, 20, 172–73, 175–76, 178, 179
Appaduri, Arjun, 128, 207

Bachin, Robin, 129
Bakhtin, Mikhail, 124
Balzac, Honore, 19, 20
Barnes, Charles, 77, 180, 183–91, 196–97
Beard, Charles, 215
Beauregard, Robert, 31
Belmont, August, 214
Benjamin, Walter, 95–96
Berman, Marshall, 236
Bloch, Ernst, 39
Boyarin, Jonathan, 19
Boyer, M. Christine, 35–36, 60–61
Bruere, Henry, 47
Buber, Martin, 30
Burnham, Daniel, 132
Byrnes, Thomas, 101

Cain, Rose, 194–95

Cartwright, Otho, 19–22, 52–55, 60
Castells, Manuel, 130
Chaterjee, Partha, 84–85
Chauncey, George, 106, 108
Cintron, Virgillio, 11–12
Clark, Cyrus, 62–63
Cleveland, Frederick, 24, 124
Cooney, Charles, 114
Cowie, Jefferson, 167
Croly, Herbert, 135
Cutting, R. Fulton, 150, 224

Daloia David, 11–12
Davis, Mike, 236
Debord, Guy, 238
DeForest, Robert, 54, 67
Delaney, John J., 148
Devery, William, 114, 161
Dowling, Frank, 222
Dyck, Isabel, 179–80, 185

Easterling, Keller, 130
Edson, Franklin, 67
Enoch, May, 113, 118
Entrikin, J. Nicholas, 124
Entrikin, Nicholas, 141

Flagg, Ernest, 69, 73, 132–34, 139, 153, 154
Foucault, Michel, 36, 126, 163

Gilfoyle, Timothy, 101–2
Glenn, John M., 185
Goldmark, Pauline, 55–60, 109, 112, 150, 151
Goodale, Henry Stirling, 157, 158
Gould, Anni, 203–4
Gould, E. R. L., 154–56
Guattari, Felix, 130

Hammack, David, 45–46, 64
Harris, Arthur, 113–14
Hartog, Hendrik, 22
Harvey, David, 27–30, 35, 44, 166–68, 178, 233, 236
Hascamp, Seth Low, 221–22
Hearst, William Randolph, 144
Hepp, John, 129
Herod, Andrew, 38–39, 168, 198, 199
Herter, Peter, 69
Herzfeld, Elsa, 154–55, 159–60
Hewitt, Abram, 71
Hills, A. A., 223
Howard, Ebenezer, 130–32
Howe, Frederic C., 122, 124, 127, 129, 132, 135, 234, 235

Irving, Lizzie, 194–95

Jacobs, Jane, 36, 137, 236
Jefferson, Thomas, 234
Johnson, Bascom, 226
Johnson, Marilynn, 97, 105, 106
Jonas, Andrew, 137
Jones, John Paul, 199

Kalb, Don, 198, 199
Katznelson, Ira, 37–38, 198
Kessler-Harris, Alice, 173, 174
Koester, Frank, 134

Lamb, Charles, 135, 138
Lardner, James, 103
Latour, Bruno, 197, 200
Lefebvre, Henri, 13, 31–35, 129, 130, 228, 236–38
Logan, John, 24, 29, 125
Lynch, Kevin, 32–33, 79

Magnusson, Warren, 162
Marsh, Benjamin, 129, 132, 135–37
Marshall, Edward, 70
Marston, Sallie, 199
Massey, Doreen, 90, 125, 168–69
McCafferty, George, 111–13
McConnon, Tom, 146–47
McCormick, Thomas, 110
McDannell, Colleen, 145
McManus, Thomas, 214, 217, 226
Minougue, Frank, 114, 118–19
Molotch, Harvey, 24, 29, 125
Mooney (priest), 204–6, 215, 217, 227, 228
Moore, Ernest C., 219
Moses, Robert, 138
Mumford, Lewis, 130

O'Hare, James, 11–12
Olmstead, Frederick Law, 135
Olmstead, John C., 135
Olsen, 149
Otter, Christopher, 50

Packard, Esther, 158
Park, Robert, 89, 90
Parkhurst, Charles, 92, 100, 144
Parsons, Frances, 149, 150
Parsons, Mrs. Henry, 121–22
Parsons, Samuel, 148
Patten, Simon, 24, 42, 127
Peete, Richard, 23
Peterson, Robert, 60

Phelps-Stokes, I. N., 54–55, 60, 73, 74, 127, 128
Piess, Kathy, 212
Plunkitt, George, 214, 217, 225
Plunkitt, William, 73, 182
Plunz, Richard, 68, 70, 153
Pratt, Charles, 153

Reppetto, Thomas, 103
Riis, Jacob, 49–50, 55, 71, 73, 209
Rockefeller, John D., 223, 224
Rodgers, Daniel, 122, 134
Roediger, David, 117
Rogers, Catherine, 155
Rogers, John, 155
Roosevelt, Theodore, 100, 110
Royce, Josiah, 124
Ruscovitch, Anton, 88

Sasken, Sassia, 137
Saxe, Martin, 220
Schneider, Henry G., 151–52
Schultz, Stanley, 20
Shivelbush, Wolgang, 234
Shklar, Judith, 40
Shrady (city coroner), 221
Shroeder, Henry, 221–22
Shurtleff, Flavel, 134–35
Simkhovitch, Mary, 207
Smith, Henry Atterbury, 155
Smith, Rogers, 41, 42
Soja, Edward, 16, 22, 30, 236

Staughton, Charles W., 222
Steffens, Lincoln, 127
Strong, William L., 66, 71
Swyngedouw, Eric, 28

Teaford, Jon, 22
Thorpe, Robert, 113–14
Tompkins, Calvin, 129, 132, 139, 180–82, 185, 197
True, Ruth, 59–60, 171–72, 212
Tuan, Yi Fu, 231

Vellier, Lawrence, 48–49, 54, 60, 67, 122, 124, 128, 139, 152
Von Ranke, Leopold, 124

Ward, David, 129
Waring, George, 63
Weatherly, U. G., 203
Weber, Max, 64
Weibe, Robert, 60
Whitman (judge), 109
Wiebe, Robert, 129
Wingate, Charles F., 72
Wood, Fernando, 98
Wood, John, 183
Woodward, Keith, 199

Yelenock, Cathy, 151
Young, Iris Marion, 178

Zunz, Oliver, 129